AN INTRODUCTION TO THE EARLY
DEVELOPMENT OF MATHEMATICS

AN INTRODUCTION TO THE EARLY DEVELOPMENT OF MATHEMATICS

MICHAEL K. J. GOODMAN

WILEY

For general information on our other products and services or for technical support, please contact our Customer Care Department within the United States at (800) 762-2974, outside the United States at (317) 572-3993 or fax (317) 572-4002.

Wiley also publishes its books in a variety of electronic formats. Some content that appears in print may not be available in electronic formats. For more information about Wiley products, visit our web site at www.wiley.com.

Library of Congress Cataloging-in-Publication Data

Goodman, Michael K. J., 1952–
An introduction to the early development of mathematics / Michael K. J. Goodman.
    pages   cm
  Includes bibliographical references and index.
  ISBN 978-1-119-10497-1 (pbk.)
1. Mathematics–History.   2. Mathematics, Ancient.   I. Title.
  QA21.G67 2016
  510.9′014–dc23

                                                                    2015036803

Set in 10/12pt Times by SPi Global, Pondicherry, India

10 9 8 7 6 5 4 3 2 1

1   2016

For Linus and Laura

CONTENTS

PREFACE

Many people simply dread mathematics. The idea that the history of mathematics might be interesting or that math problems could be solved as a recreation astonishes them.

I like math. When I see a mathematical idea for the first time, I wonder: *Where did that come from? Who came up with that, and why?* And my effort to answer these questions often leads to a pleasant and intellectually exciting investigation of books and websites. Some of my students shrug anyway, and say: *Who cares?* I tell them why they should care. Every idea that expands their minds makes them better people, more capable of doing the tasks they have chosen to do, and more interesting to talk to during lunch.

The history of any subject is a worthwhile study, and while art, science, music, and medicine have long been taught at colleges, in recent decades there has been a proliferation of courses on the history of art, the history of science, the history of music, and the history of medicine. The history of mathematics is enjoying a similar surge in popularity, although perhaps not for the best of reasons. Just as you can have no artistic talent yourself but still enjoy and excel at a course in the history of art, or be inept at scientific work but still excel at a course in the history of science, you might find comfort in the history of math despite finding no comfort at all in math itself. Many of my students have stumbled into my classroom just because they were scared to take calculus or statistics.

I see the history of math as closely allied to the field of recreational mathematics. In recreational math, we solve problems not because they are in a textbook, but because they are part of a story that contains a mathematical puzzle. Recreational math is about solving brain-teaser-type problems in a clever way. It is like solving crossword puzzles, and some of the newer puzzles that are in vogue, like kenken and sudoku. It's problem solving without the drudgery, and finding a solution gives the solver a feeling of satisfaction and success. In a college course in the history of mathematics, there are a lot of recreational math problems to solve, using clever techniques which, when properly applied, often lead to a feeling of satisfaction and success. The ancient Egyptians, Chinese, Babylonians, Greeks, Hindus, and Arabs had a wide range of mathematical interests, and there are plenty of concepts and calculations to discover and admire, and even to understand and emulate. When we try to do math their way, we are doing it primarily as a game, not for practical reasons.

Many of my students have said that they wanted a small book that summarized the historical background behind the mathematical accomplishments of ancient civilizations and also explained how to do representative math problems, and that is why I wrote this book. It is intended to do exactly those two things, briefly to give a bit of the history and to show solutions to typical problems, so that the students would have a guide to fall back on when they themselves tried to do Babylonian multiplication or Egyptian fractions or any similar operation based on ancient methods.

In many ways, this is a "how to" book. How to solve a Chinese problem in a Chinese way. How to solve an Arabian problem in an Arabian way. It is also a "why" book. Why did the Greeks make such astonishing progress in mathematics? Why did the Hindus want to calculate the square root of 2 so precisely? It is also a "proof" book. I will show a proof that verifies that an Egyptian algebra technique works. I will show a proof that verifies that a Babylonian geometry technique works. But mostly, I hope that you find this an interesting "story" book, which ties together the hows and whys and the proofs in a way that makes the history of mathematics a pleasant subject to think about and want to know more about.

I have a great deal of enthusiasm for the history of mathematics, partly because I like to tinker with numbers. The study of ancient math makes it clear that other people were enjoying tinkering with numbers thousands of years ago. Yes, mathematics developed because there were practical problems to solve in agriculture, government, and war, but once the early mathematicians had worked out the math they needed for their business, they apparently played around with numbers a lot. So many of the problems preserved in ancient writing are not practical at all—they are frivolous, gimmicky, and recreational, meant to entertain and let the cleverest mathematicians show off a little bit.

There have been a few times in my lifetime of tinkering with numbers that I have "discovered" something wonderful and surprising, and I thought I had an original result that the mathematical community would be thrilled to know about. But when I researched the history of each topic, I found that others had found these wonderful and surprising results long ago. My insight into the sums of squares and cubes was known by the Arabs 1000 years ago. My insight into a different way to define a circle was known by the Greeks 2000 years ago. My insight into a way to find right triangles was outdone by the Babylonians 3000 years ago because they had a way to find even more right triangles than I had found.

It would be greatly satisfying to me to know that this book had in some way inspired you to tinker with numbers. More likely though, this book is a means for you to pass some time or pass a course, and I can be somewhat satisfied if the book entertained or taught you something and caused you to be curious about ancient people or appreciative of their skills or perhaps in awe of their skills—they did this mathematical stuff without calculators and computers.

Chapters 3–5 and 7–9 in this book cover events in the history of mathematics in six important civilizations: Egyptian, ancient Chinese, Babylonian, Greek, Hindu, and Arabian. They are presented in a rough chronological order, related to when these civilizations made significant mathematical progress (which we know about!), but the chronology is very rough, and the reader should feel absolute freedom to acquire the information in any different sequence. In fact, historians will tell you that the earliest cultural activities that we recognize as essential to a "civilization" occurred in Mesopotamia, over 5000 years ago, under the direction of people we know as Sumerians. Within a few hundred years, Sumerian control had given way to Babylonian control, and via outstanding archeological good fortune, we know a lot about Babylonian mathematics.

I present Egyptian mathematics first. Egyptian civilization is nearly as old as Mesopotamian civilization, and Egyptian math is much easier than Babylonian math, a key consideration when your audience is new to the overall subject.

Shortly after the Egyptians had mastered their environment, another civilization sprang up in the region of the Indus River; but in this case, there is no archeological good fortune to speak of, and while the remains of cities assure us that these people were well acquainted with mathematics—you hardly get architecture and engineering without math—there is little to be said about this civilization. Instead, the descendants of these people, the Hindus, are treated in a later chapter.

After covering Egyptian math, I turn to Chinese math. China was one of four civilizations that emerged in the next time interval, still over 4000 years ago, but unlike the other three, its mathematics is well-documented. The Elamites in Iran and the Hittites in Turkey left us nothing particular in math (their work may have been absorbed by the Babylonians), and the Minoans in Crete are an enigma, although eventually their ideas may have influenced Greece. Even to say China's math is well-documented is something of an exaggeration: Chinese civilization flourished for thousands of years, and the mathematical history we know comes mostly from the middle and later periods, not the earliest. Of course, the same points about longevity and documentation apply to Egypt and Babylonia also.

Next I present Chapter 5 on Babylonian math. It simply cannot be put off any longer, despite its thoroughly different approach to representing numbers.

The math of classical Greece comes next. The Greeks undoubtedly benefited from the work of the Egyptians and the Babylonians who came before them. They may have also interacted advantageously with the Hindus, who were their contemporaries, although geographically distant. The final group to be discussed is the Arabs, who built their mathematics on a foundation laid down in both Greece and India.

I decided to limit the content of this book to events that occurred before AD 1000. The volume of math that comes after this date dwarfs the volume of math that comes before it, and my intent is to present the right amount for a one semester introductory course.

For many students, the most intimidating aspect of the study of ancient math is the unfamiliar number systems, but this obstacle is easy to overcome and curiously not an essential element to many parts of the story. We can do multiplication the Egyptian way either with their hieroglyphic numbers or with our everyday numerals, so understanding their method is more important than knowing their penmanship. The Internet is a fantastic resource for seeing all the old number systems, and with sufficient patient practice, any student can replicate the symbols well enough.

Chapter 3 includes a few illustrations with Egyptian hieroglyphics, but most of the work is done with our numbers. The work includes two different Egyptian techniques for multiplication, their technique for division, their special way of representing fractions, and their special way of solving basic algebraic equations. A good deal of what we know about ancient Egyptian math comes from problems preserved on papyri, so we look at a variety of these problems.

As in all the other major chapters, the student is challenged with "problems" and "exercises". *Problems* require calculations, and they echo sample problems presented in step-by-step detail in the text. Answers to odd-numbered problems and solution techniques appear at the end of each chapter. *Exercises* generally require extended thought and can safely be ignored by a student who is not particularly curious. For the curious, a collection of suggestions and solutions appears in Appendix A.

Despite the aforementioned chronological facts, the chapter on ancient Chinese mathematics comes before the chapter on Babylonian mathematics because in my experience every student finds Chinese math easier than Babylonian math, not that either one is easy. Naturally, the starting point with Chinese math is the calligraphic number system, which is different in substantial ways from all the number systems a student is likely to know: our decimal system (using Arabic numerals), the Roman numerals he or she learned as a child, and the Egyptian numbers from the previous chapter. The key feature of this system is a multiplicative element, where one character tells you by how much to multiply another character.

After a discussion of how the Chinese represented numbers and did basic arithmetic, the focus of the chapter is problem solving. Chinese records include a rich selection of algebra and geometry problems dressed up in fanciful stories. Sometimes, the premise seems absurd, like a farmer needing to buy exactly 100 animals for exactly 100 coins, but the stories seem intended to guide aspiring students to competently handle systems of equations, rates, similar triangles, areas and volumes, and division and remainders. Problems range from constructing simple magic squares to factoring huge numbers that arise in equations with variables raised to the fourth power. An especially interesting category involves a branch of advanced arithmetic called "number theory" and a concept that we know today as the Chinese Remainder Theorem.

Since the Babylonian cuneiform number system is so different, a great deal of Chapter 5 is devoted to to the ins and outs of base 60 arithmetic. Number representation, place value, addition, subtraction, multiplication, division, and fractions are explained. Since the symbols for the Babylonian digits are difficult to draw, an expedient compromise has been made that helps both me and the student: the digits are shown as simple triangles rather than drawn more subtly as they appear on the thousands of hardened clay tablets that today are testimony to ancient Babylonian brilliance. I highly recommend visiting internet sites that show the actual tablets, and two of the videos I recommend (in a section listing Recommended Websites and Videos after Further Reading) show modern researchers patiently etching cuneiform wedges into soft clay.

The scope of Babylonian math is broad, so only a few really exceptional topics are covered: a dazzling calculation involving the area of a trapezoid, a technique for solving quadratic equations, formulas for square roots and cube roots, and an astonishing list of triangles that conform to the Pythagorean theorem, etched into a clay tablet many centuries before Pythagoras was born. A great enduring mystery about the Babylonians is why they recorded their mathematics in the base 60 format. Some speculations are discussed in the chapter, and in Appendix B I present my own theory on the subject.

The chapter on Greek mathematics is a scratch of the surface. Whole books have been written about Greek math, and in fact whole books have been written about individual Greek mathematicians, like Archimedes and Euclid. We have dozens of ancient Greek math textbooks fully preserved and fragments or records of the contents of many others. Today's students hardly realize it, but to a surprising extent they are already experts in Greek mathematics, since virtually the entire math curriculum from middle school to pre-calculus is based on the work of Greek geniuses. Some brilliant Italians gave us negative numbers early in the Renaissance, some clever Frenchmen gave us probability and analytical geometry during the Enlightenment, and some Germans and Englishmen blessed us with integrals, derivatives, and logarithms, but outside of that it's pretty much all Greek. We'll see what Thales discovered about triangles and circles, what Pythagoras knew about triangular numbers and angles in polygons, what Hippasus found out about the square root of 2 that got him in trouble, how Hippocrates almost found a square and a circle with the same area, what Democritus calculated about cones and pyramids, the 5 shapes that Plato thought were perfect, how Euclid proved there were more prime numbers than we could ever count, how Archimedes estimated pi, and on and on and on.

Chapter 8 concentrates on the entertaining story problems Hindu mathematicians created to challenge each other and dazzle their patrons. Typical problems involve the price of fruit, the path of a flying wizard, and the length of an ever-growing snake. To solve these problems, the Hindus, like the Chinese, relied on systems of simultaneous equations, related rates, and the Pythagorean theorem. In one problem, where an exchange of gifts leads to a surprising coincidence, the solution is essentially a fundamental principle of modern matrix algebra. Influenced almost certainly by the Babylonians, the Hindus improved on the calculations of square roots and cube roots. Influenced by the Greeks, they made advances in number theory. The Hindus had a remarkable geometric method for very accurately computing the square root of 2.

Chapter 9, describing Arabian math, is our bridge to the modern world. After the fall of the Roman Empire, Europe lost both the means and the desire to preserve and expand the mathematical insights of the classical Greeks. The common term we learn in history class for this period is the "Dark Ages," and as far as European culture is concerned it is a reasonable term. However, not far away, in the countries controlled by the Arabs, the situation was completely different. Greek mathematical manuscripts were devoured by scholars anxious to add their ideas to the Greek foundation. So, while in Europe a spectacular treatise by Aristotle was erased so that a monk could re-use the pages to write down some prayers, in Arabian libraries Greek texts were being copied, translated and annotated. (A wonderful video about Archimedes' lost, and subsequently rediscovered, book is cited on page 230.)

The Arabs busily solved second and third-degree equations, examined geometry and trigonometry in detail, experimented with number theory, and applied the practical things they discovered to the increasingly important fields of astronomy and accounting. Astronomy was vital for regulating the calendar, for navigation, and for the Muslim's obligation to face Mecca while praying. Accounting was useful for the growing economy and the administration of an ever-increasing territory. What the Arabs preserved and created was eventually transmitted across the Mediterranean to the Venetians, Florentines, and other politically prominent inhabitants of the Italian peninsula.

It will undoubtedly occur to many readers that *something* must have happened in mathematics before the Sumerians and Egyptians began their clever applications of it, and that is the subject of Chapter 2, a short and somewhat speculative chapter. The field of mathematical anthropology is wide open for research and theorizing, and it can be approached from a few distinct directions. First, discovery of new anthropological sites, or more complete interpretation of known sites, could clarify mankind's earliest mathematical thinking. Second, it is very plausible that the remote and isolated populations around the globe still living in Stone Age societies might have the same understanding and use of numbers that our remote ancestors had. Third, other fields, including neuroscience and linguistics, might delimit dramatically what early humans were capable of.

And it is also quite likely that some readers will wonder how it is that we know what we do happen to know about the mathematical accomplishments of ancient civilizations, and this is largely a matter of archeology. Curious scientists have dug up the old cities of Egypt and Mesopotamia and have translated the words written in long dead languages. In Chapter 6, devoted to mathematical archeology, the challenges of interpreting artifacts are described, and the reader is presented several opportunities to step into the role of an archeologist, on the job, trying to make sense of a mathematical puzzle. Chapter 6 also features brief discussions of Roman numerals and Mayan numerals.

In the Introduction, I'll address a few words to the student using this book in a class, based on my experience in teaching a class called *The Development of Mathematical Thought* at Westchester Community College in New York. Those words are intended to be encouraging and are in an introduction because students generally do not read prefaces, not being especially concerned about why a book was written or how a book was written.

MICHAEL K. J. GOODMAN

ACKNOWLEDGMENTS

Producing a book such as this one is a collaborative effort. John Teall, who has written his own textbooks, was a key advisor. Stan Wakefield an expert on all phases of the book business, brought my manuscript to Wiley. The Wiley editorial and production teams guided me through the process that turns a manuscript into a book. Particularly helpful and encouraging were Susanne Steitz-Filler, Sari Friedman, Divya Narayanan, Nomita Swaminathan and Shiji Sreejish.

There are many fine math professors at Westchester Community College. The first time I taught a course on the history of mathematics, Sean Simpson gave me his notes and class handouts. Jodi Cotten, who teaches a course called "The Nature of Mathematics," made many useful comments about our shared interests. Joyce Cassidy, who teaches Geometry, presented the well-known core of that subject in a thoroughly original way. And Ray Houston, who also teaches the history of mathematics, has frequently put me on the schedule to teach my favorite course.

I had the good fortune, when I was a student, to know some real mathematical geniuses, and one of them, Jonathan Watson, has kicked mathematical ideas around with me for 50 years. Other friends, who have made things possible or made things easier or made things wonderful are Andrew Goodman, Juan Leon, Olivia Masry, Alan Koblick, and Anne Teall.

And my wife Franziska and my daughter Emily contributed to this book in several ways. Franziska knows a lot about preparing a manuscript and taking advantage of the options and subtleties of software, and Emily had the courage to take a very advanced college course in number theory and the graciousness to tell me all about it.

1

INTRODUCTION

This book is not a comprehensive history of mathematics. There are excellent books, much longer than this one, that present a thorough survey of ancient math. There are also more specialized books that go into great depth about certain topics (the history of algebra, the development of numerals, etc.) or certain civilizations (math in ancient Greece, math in ancient China, etc.). I hope that you are interested to look at these other books one day, and some of them surely are in your college library right now.

When I start my course, I tell the students that this is obviously a math course. It's offered within the Math Department, and the students will solve math problems. Some of the problems are practical, and some of the problems are exercises that students had to do in their schools in Egypt and Babylonia long, long ago. It's a *math class*.

It's also a *history class* because what we're exploring here is where ancient mathematical ideas came from, what kinds of problems these old civilizations needed to solve, and how mastering math helped them develop their agriculture, engineering, government, economic, military, and social systems. So it's a *history class*.

I tell my students it's also a bit of an *art class*. Not everybody could read and write in the ancient world. In fact, it was a select minority who could. And a long period of training and education was required to train the scribes who did the counting and computing and administrative work that made ancient Egypt, ancient Babylonia, and ancient China run. I have my students practice making the numeric symbols that these old civilizations used, just as if they were in a scribe school, and this takes patience and accuracy.

An Introduction to the Early Development of Mathematics, First Edition. Michael K. J. Goodman.
© 2016 John Wiley & Sons, Inc. Published 2016 by John Wiley & Sons, Inc.

It's also a bit of an *archeology class* and an *anthropology class*. If we're talking about the mathematics used by old civilizations, we need to know a little bit about these civilizations, and generally students get curious about how we know what we know about these civilizations—that's where the archeology comes in. And if we go further back in history, to the nomadic, tribal, primitive kinds of societies that lived before the agricultural revolution, we get into the realm of anthropology—how and why did people even begin to think mathematically? Remarkably, there are groups of people around the world who were so isolated until modern times that we can use what we know about them to make reasonable guesses about our quite remote human ancestors.

I even tell my students this is a bit of an *English class*, because I ask them to write a short paper about a mathematician or about a bit of mathematics. Basically I'm looking for one idea that can be linked to one mathematician. Students can choose something from long ago (like Archimedes working out an approximation of π) or something modern (like Mandelbrot developing fractals) or something in between (like Omar Khayyam solving cubic equations). I look for clear writing that explains the idea, describes the world the mathematician lived in, and relates the idea to mathematical knowledge that came before it or grew out of it.

So, don't expect everything from this slender book. It is a starting point. I have my own favorite websites about the history of math, but your professor may have his (or hers) and there are many good ones. I show my students videos of people making cuneiform tablets and of people building models of Platonic solids. I show the Mayan codex books that survived and I also show the Hindu derivation of the square root of 2. I read out loud to the class the view of the Crow Indians that honest people don't need numbers larger than 1000.

So I urge you to read, think, and be curious. You've got an electronic calculator at your fingertips and the advantage of a clever base-10 system with 10 simple digits. But put yourself in the position of someone thousands of years ago, who needed to solve an arithmetic problem and didn't have these wonderful and convenient things. Put yourself in that position often enough this semester, and you'll have a profound understanding of the history of math and the development of mathematical thought.

2

MATHEMATICAL ANTHROPOLOGY

In the chapters to come, this book will focus on major ancient civilizations and the development of math in those civilizations. You probably already anticipate chapters on Egypt, Babylonia, and China. But it is natural to wonder: what came before that? Where did the very earliest mathematical ideas come from?

It turns out we have some information. Anthropologists have found, during the past few centuries, dozens of isolated societies that live (or lived) with Stone Age technology. These societies were found in remote areas, like Pacific islands and tropical jungles. These societies got along perfectly well, for thousands of years, without contact with the rest of the world.

From the study of these societies, anthropologists are able to make an informed guess about the lives of very early humans. We might think of them as *cavemen* or *hunter-gatherers*, or *primitive people*, but whatever label we give them we have to recognize a few key things about them:

- **They were around for a very long time**. The paleontological evidence shows humans with our brains and our capacities were living for tens of thousands of years before recorded history.

- **They included an enormous range of cultures, all different from each other**. The aborigines of Australia included people who spoke very different and mutually unintelligible languages. These aborigines were quite unlike their contemporaries in New Guinea and Tasmania, and even more unlike their contemporaries in Indonesia, Micronesia, and southeast Asia. All these people

An Introduction to the Early Development of Mathematics, First Edition. Michael K. J. Goodman.
© 2016 John Wiley & Sons, Inc. Published 2016 by John Wiley & Sons, Inc.

had decidedly different lifestyles and cultures than the people who lived in the rest of Asia, in Africa, Europe, the Americas, and the Arctic.

- **They knew enough to thrive in their varied environments**. They did not starve, die from diseases, get completely killed off by other tribes, or go extinct because of any of the hazards they faced. They survived.

Part of what kept them alive is that they could count.

There were a lot of things to count. *Are we all here?* is an important question. *Do we have all our stuff?* is another. *How many fish do we need to catch? How many days will it take to walk to the next place? How many people are in that band of strangers you saw?* (If we outnumber them, they probably are not a threat to us.)

The ability to count and communicate numbers gave a group a big advantage over a group that couldn't count and communicate.

The earliest evidence of counting is notched bones and sticks. One famous bone, from Africa about 20,000 years ago, appears to have been cut so that the person holding it could use the notches to count something. Was he counting animals, people, days, or something else? That, we don't know. But the regular, evenly spaced notches suggest that this tool was used again and again for the purpose of counting. The clusters of notches come in intriguing quantities (like 11, 13, 17, and 19) and have engendered much analysis and speculation about what early people may have known about numbers.

It also makes sense that some fairly large number had to be counted. All those primitive societies anthropologists found on islands and in jungles counted on their fingers (and sometimes toes and other body parts if fingers weren't enough). There would be no need to cut dozens of notches into a bone (and perhaps carry the bone around from place to place) if the number of things that had to be counted was only a small number. So the counting bone was probably a very important possession.

Since we are speculating about prehistory, we cannot know just how and when certain mathematical ideas arose, and whether they came all together or one by one. We can't be certain how long it took to develop an idea, and the order in which the ideas came. What we can say with some confidence is that these ideas were discovered again and again, in different bands and tribes and societies, because of the isolation of people. It is likely that the world's population 10,000–30,000 years ago consisted of a very large number of very small groups who knew only the people who lived quite near them or along their migratory routes. The exchange of ideas in these circumstances would have been limited.

Here are some of the ideas that must have been discovered and developed countless times.

COUNTING

This is natural. It is advantageous to count. What might be surprising is how many ways there are to count. Many primitive groups anthropologists studied had specific names for numbers up to a certain maximum, and then named larger numbers as combinations of the smaller ones or just said "many" and did not distinguish among the larger numbers at all.

At one extreme, the Semang people of the Malay peninsula had no number larger than 3. So their system was 1, 2, 3, many. Tasmanians could count as high as 5. A group of polar Eskimos were said to have numbers from 1 to 5, but they represented higher numbers as combinations: 6 was "5 and 1" and 10 was "two 5s." An aborigine group counted thus: 1, 2, "one 1 and one 2," "two 2s," and so on.

Quite a few primitive groups, including the Dahomeans of west Africa, gave special attention to 5 (the fingers on one hand) and 20 (all the fingers and all the toes together) in their figuring. Among the Ainu in northern Japan, 800 was the highest recognized number. 800 is 40 groups of 20.

The vast majority of languages show that (i) counting was influenced by our having 10 fingers, and (ii) words for smaller numbers can be combined to make words for larger numbers. (Isn't our word *eighteen* a reflection of our words *eight* and *ten*?)

ORDER DOESN'T MATTER

The fellow who was using a notched counting stick to see if all the people in his group were present didn't have to count them in any special order. It may have been easy to remember them in a certain order (adults first, then children, immediate relatives before clansmen), but simply matching a notch to a person allowed for a quick count.

THE COMMUTATIVE PROPERTY

If you're counting something, like axes, it doesn't matter whether you count the ones in this pile first and then the ones in that pile—you get the same total regardless of which pile you count first.

ADDITION AND SUBTRACTION

If you make a new axe, you have more. If you lose an axe, you have less.

ONE-TO-ONE CORRESPONDENCE

Sometimes, you really need to have the same number of two things. It's not the most efficient hunt if you send 6 hunters out with 5 spears. Even more critically, for harmony within the group you may want exactly the same number of adult men and adult women.

RECURRING PATTERNS

The moon was pretty important in those days, because you could see so much more on a night with a full moon. Somebody must have noticed that those moonlit nights came on a reliably regular schedule.

The Haida Indians of British Columbia, the Witoto tribe of northern Amazonia, and the Ganda of eastern Africa, among others, based their calendars on the cycles of the moon.

GREATER THAN AND LESS THAN

Recognizing that some numbers were bigger than others was extremely useful. A group probably had little to fear from a group smaller than itself, but it might have to flee for safety when a larger group arrived.

Curiously, primatologists who have studied aggressive chimpanzees concluded that chimps have an outstanding ability to recognize greater numbers: they only attack when they have a numerical advantage, and chimpanzee "wars" can go on for years.

FRACTIONS (COMPARED TO WHOLE NUMBERS)

Sharing food probably forced primitive people to confront the idea that some things were divisible.

The Incas of Peru had a decimal (based on 10) system and routinely required smaller units of their empire to provide $\frac{1}{10}$ of the labor or $\frac{1}{10}$ of the taxes.

DOUBLING, HALVING

For many tasks, two heads were better than one, or two able bodies were better than one. Who wouldn't have noticed that the fingers on two hands were double the number of fingers on one hand? When it came to sharing food or anything else, wasn't cutting in half the fairest distribution?

RATES

Some sense of how long it took to walk from point A to point B could give you a way to describe other distances and other durations.

RATIOS

If a bark canoe could hold 3 people and their possessions, you needed to build 4 canoes to transport 12 people and their possessions.

BIG NUMBERS

The Toda of southern India could confidently count into the thousands. The Iroquois of New York into the hundreds of thousands. And the Ganda (Africa) into the millions.

Undoubtedly, this is a very simplified and potentially incorrect snapshot of early mathematical ideas, but by the time there are written records of math, the math has progressed far beyond what I've just listed. So we have to believe that the accumulation of mathematical wisdom was slow, incremental, universal, and poised to explode when the agricultural revolution made new mathematical insights possible and necessary.

3

ANCIENT EGYPTIAN MATHEMATICS

It's amazing how much we know about ancient Egyptian mathematics. There is probably also a great deal we do not know, considering how many centuries have passed, but we can say we know the kinds of math problems the ancient Egyptians worked on and the techniques that they used. We know their hieroglyphic number system, we know how they multiplied and divided, we know their fascination with a certain kind of fraction, we know how they solved algebra problems, and we know how they trained students to master the math necessary to make government and business run smoothly.

A lot of what we can say about the role of math and development of math in Egypt could also be said, with modest modification, about ancient Babylonia, ancient China, and ancient India. These are the sites of the great river valley civilizations where humans made the great transition from living in small nomadic or seminomadic bands to living in large and permanently settled communities. The big difference was agriculture. Wild grasses and wild animals were tamed, and well-situated farms led to reliable supplies of food. Good climate and plentiful water were available along the banks of Nile River (Egypt), the Tigris and Euphrates rivers (Mesopotamia), the Indus River (actually in Pakistan now), and the Yellow river (China).

Agriculture required ever-improving measurement and mathematics. A farmer had to know where his land ended and the next farmer's land began. He had to know how many seeds he needed to plant, how many animals he had to feed, and how many workers could do how much work. If he wasn't right alongside the Nile, there was first the question of how much water could be carried to his fields and then later the

An Introduction to the Early Development of Mathematics, First Edition. Michael K. J. Goodman.
© 2016 John Wiley & Sons, Inc. Published 2016 by John Wiley & Sons, Inc.

geometrical calculation of where to build an irrigation ditch and perhaps even later a canal. Surplus food was stored collectively for the future, and an accurate account of who had contributed how much was needed.

The cyclic regularities of the seasons and the weather and the flooding of the Nile made calculating the length of the year important. This tied directly to observations of the positions of the sun and the stars, and of course the movements of the moon and the planets were noticed. Astronomy led to religion, and in ancient Egypt there was no distinction between church and state, as the Pharaoh, the king, was also a god. So agriculture, astronomy, religion, and government were all connected, and mathematics was a vital servant of them all. There were taxes to collect, temples to build, armies to equip, and cities to administer. When you think of ancient Egypt, probably the first thing you think about is the pyramids. Imagine the math problems to be solved to design the pyramids (architecture) and to build them (engineering).

And while we take for granted today that education is universal and anyone can rise through talent and hard work, these modern ideas do not apply to ancient societies. Education, and even literacy, was reserved for an aristocratic minority. To become a vital member of the powerful government bureaucracy, a boy (it was only the boys then) had to endure many years of strict and demanding education, mastering the hieroglyphic writing system, the number system, the mathematical operations, the laws and customs of society, the geography and history of Egypt, and even foreign languages.

But the rewards for attaining the office of scribe or bureaucrat were enormous. For an honest man, there was status and prestige and certainly some measure of wealth. For a dishonest man, there were some great temptations and opportunities. Imagine the advantages a trader or tax collector would have if he could read and do arithmetic and he knew the people he dealt with could not.

Let's look at the Egyptian representation of numbers. The system is simple. There is a symbol for 1, and a symbol for 10, and a symbol for 100, and for each succeeding power of 10. The symbols were repeated for multiples: they wrote two 10s for 20, three 10s for 30, and so on. 5678 would be written as five of the symbols for 1000 (a lotus flower, as illustrated in Fig. 3.1), followed by six of the symbols for 100 (a coiled rope), and seven of the symbols for 10 (a heel bone, which looks something like a horseshoe), and eight 1s (vertical strokes just like ours). The symbols might be written left-to-right (the way we read), or right-to-left, and might be written as a long row or in bunches of short vertical stacks.

There was no symbol for zero, but they got along without it. There was a symbol for fractions, so 8 was distinguished from $\frac{1}{8}$. Adding and subtracting were straightforward, but multiplication and division were generally based on doubling numbers

FIGURE 3.1   The number 5678 represented in Egyptian hieroglyphics.

and cutting numbers in half. Equations were solved via a guessing technique, and fractions were expressed in a special way, where the numerator was 1: they would write a sum, $\frac{1}{2} + \frac{1}{4}$, where we would simply say $\frac{3}{4}$. Fractions with a numerator of 1 are called unit fractions, and the Egyptians found ways to express all fractions as the sum of unit fractions.

There have been criticisms of Egyptian mathematics. The Egyptians were not as advanced as the Babylonians. They were not as systematic as the Greeks. They estimated π less accurately than other people did. They didn't use the Pythagorean theorem. They handled math problems as unique cases, never generalizing and finding laws, and regularities that other people did.

But, for all these criticisms, it can be argued that Egyptian civilization lasted for thousands of years, and they built the pyramids, and maybe because we have only such a small portion of their work, we just don't have any record of the laws and regularities and generalizations they knew. Maybe.

Let's look at the doubling ideas behind Egyptian multiplication and division.

A simple problem is 35×12. An Egyptian scribe would recognize that the answer is the sum of a certain number of 35s and would start doubling and redoubling 35, in columns like this:

1	35
2	70
4	140
8	280

You could read this table as "one 35 is 35, two 35s are 70, four 35s are 140, and eight 35s are 280." The scribe would know this table is big enough to solve the problem because the next number in the doubling sequence (1, 2, 4, 8) is 16, which is more than 12, and he's multiplying 35 by 12.

So he'd look for numbers that add up to 12 (4 and 8), and add the multiples of 35 that correspond to these rows of his table (140 and 280). The sum of 140 and 280 is 420, which is the answer to 35×12.

Now, of course, the Egyptian scribe would not write the numbers the way we write them. His work would look like this:

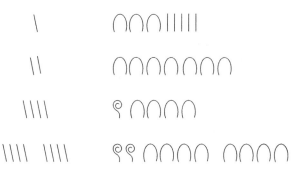

FIGURE 3.2   An Egyptian scribe's step-by-step work on a multiplication problem.

On the left side, you see the number of vertical strokes doubling from line to line, from 1 to 8. On the right side, you see on the first row 3 heel bones and 5 vertical strokes, the Egyptian representation of 35. The second line has 7 heel bones, which shows doubling and a conversion of symbols: if you start with 3 heel bones and 5 vertical strokes and then double these symbols, you get 6 heel bones and 10 vertical strokes; the scribe would know that 10 vertical strokes should be replaced by 1 heel bone, and the conversion the scribe would do is essentially what we do when we say $60+10=70$.

The third row shows 70 being doubled to 140; 7 heel bones doubled is 14 heel bones, but the scribe would also know of course that 10 heel bones (10×10) can be replaced by 1 coiled rope (1×100).

The fourth row is easy. The numbers of coiled ropes and heel bones are simply double the numbers seen on the third row.

The scribe would see that 4 vertical strokes and 8 vertical strokes (on the left side) together made 12 vertical strokes, so he would add the coiled ropes and heel bones on the opposite side. He'd get 3 coiled ropes and 12 heel bones, but then he'd convert 10 of the heel bones to 1 more coiled rope, ending with 4 coiled ropes and 2 heel bones (420).

There was a second multiplication technique, where one factor was doubled successively and the other factor was cut in half successively. That would generate a table as follows:

12	35
6	70
3	140
1	280

Students always wonder about cutting odd numbers in half. Half of 3 is 1.5, not 1, but the table only shows whole numbers. Well, the Egyptians threw the fractional part away when they used this technique.

The next step is to add the numbers in the second column that are opposite *odd numbers* in the first column. This gives us $140+280$, which we saw before was the right answer.

So, the Egyptians definitely knew which numbers were odd and which were even, and they had two perfectly good ways to multiply whole numbers. A good student exercise is to demonstrate the reliability of the two Egyptian multiplication methods.

EXERCISE

Confirm that every positive whole number can be expressed as the sum of some combination of the powers of two (1, 2, 4, 8, 16, 32, 64, 128, 256, etc.) without using any power of two more than once. This verifies the first Egyptian method.

EXERCISE

Confirm that the second Egyptian method produces the same results as the first method, particularly with regard to which numbers to add to find the final result. This is a somewhat harder exercise.

EXERCISE

Confirm that the results are the same whether we compute twelve 35s or thirty-five 12s by the first method, and whether we swap which number gets doubled and which gets cut in half in the second method.

PROBLEM 3.1 Use the Egyptian doubling method to multiply 67×42.

PROBLEM 3.2 Use the Egyptian doubling method to multiply 33×99.

PROBLEM 3.3 Multiply 68×43 using the Egyptian method that combines doubling and dividing by 2. (The name for this method is "duplication and mediation.")

PROBLEM 3.4 Multiply 75×49 with the "duplication and mediation" method.

So far, we haven't been concerned with numbers that are more than a few thousand, but the Egyptians wrote larger numbers easily. Figure 3.3 shows the standard symbols for the powers of 10, up to 10 million.

The symbol for 10 million is apparently the rising sun. The symbol for 1 million is an astonished man, seated but raising his arms. The symbol for 100,000 is a turbot fish, a fish with its long tail bent at a broad angle, perhaps leaping rather than swimming. The symbol for 10,000 is a pointing finger. (Some people think it is a snake, not a finger. Whatever it is, it's a long, thin shape.) We don't know why these

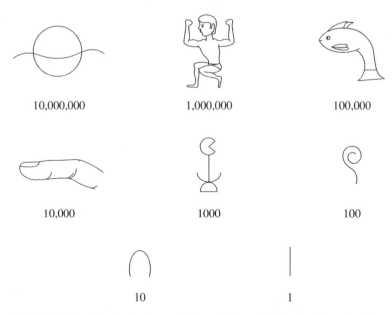

10,000,000   1,000,000   100,000

10,000   1000   100

10   1

FIGURE 3.3 The standard numerical symbols use in Egyptian hieroglyphics.

particular pictures were chosen, but it seems reasonable that the sun would represent a bigger number than a man, since in the Egyptian religion the Sun was a god of great importance. Similarly, a man (1,000,000) would have higher status than a fish (100,000) or a mere part of man (10,000), and we can guess that flowers, ropes, bones, and simple lines had even less status. Also, since the smaller numbers were used more frequently, it would have been convenient to make the symbols for smaller numbers easier to draw.

Now, there are two comments worth making about Egyptian number symbols. First, these symbols evolved over time. Some symbols were just drawn differently as the centuries passed. Think about looking at the familiar letters you are looking at right now, compared to looking at documents from America's colonial period. The old-fashioned colonial lettering is different—we can read it, but it's different. Some S's in our Declaration of Independence look like extra long F's, for example. Well, over the centuries, something similar happened with Egyptian symbols. In one particular case, the thing that was being drawn even changed: the fish that stood for 100,000 was replaced with a tadpole. In some Egyptian drawings, the vertical stroke for 1 and the heel bone for 10 are depicted as short lengths of rope, apparently fractions of the coiled rope that represented 100.

The second comment is that the numbers we are looking at are one particular style, the *hieroglyphic numbers*. A second style, the *hieratic numbers*, using a completely different set of symbols, was developed later, and it was a lot easier to do and record arithmetic with this style.

Some students enjoy translating our numbers into hieroglyphics and translating hieroglyphic numbers into our familiar numbers. Some don't particularly enjoy it, but find that the exercise gives them an understanding of how tedious the work of a scribe could be. One thing to observe is that the number of symbols is not particularly well correlated with the size of the number. For example, writing 97 requires 16 symbols (9 heel bones and 7 vertical strokes), while writing the vastly bigger 301,020 requires only 6 symbols (3 fish, 1 lotus flower, and 2 heel bones). You will notice that for a number like 301,020 the Egyptians don't insert zeroes (or their equivalents) the way we do; the absence of pointing fingers, coils of rope, and vertical strokes in 301,020 implies the zeroes.

For division, the Egyptians more or less reversed their multiplication technique. Since we just saw that $35 \times 12 = 420$, we know that 420 divided by 35 is 12.

The Egyptians would make their reliable doubling table to start.

1	35
2	70
4	140
8	280
16	560

They could stop at 560, since it is more than 420. (They could even stop at 280, since 280 is more than half of 420, but that requires an extra mental calculation.) Once we have doubled our way to a sufficiently large number, the idea is to subtract the biggest number on the list that is less than (or equal to) the starting number. Since we're dividing 420, the biggest number on the list less than 420 is 280.

420 − 280 = 140. Now we subtract the biggest number on the list less than (or equal to) 140, which is 140. The 280 was opposite the 8, and the 140 is opposite the 4, and we add those two numbers to get our answer, 12. The work looks like this:

$$
\begin{array}{r}
420 \\
280 \\
\hline
140 \\
140 \\
\hline
0
\end{array}
$$

The interpretation is: 420 is made up of a certain number of 35s. We found that 420 is composed of 280 (eight 35s) and 140 (four 35s), so 420 is composed of twelve 35s.

Let's do this with a result we don't already know: 1000 divided by 47.

The successive doubling of 47 gives

1	47
2	94
4	188
8	376
16	752
32	1504

We can stop because 1504 is more than 1000. The biggest number on the list smaller than 1000 is 752, so we subtract

$$
\begin{array}{r}
1000 \\
752 \\
\hline
248
\end{array}
$$

Now we're looking for a number on the list less than 248.

$$
\begin{array}{r}
248 \\
188 \\
\hline
60
\end{array}
$$

Now we're looking for a number on the list less than 60.

$$
\begin{array}{r}
60 \\
47 \\
\hline
13
\end{array}
$$

There are no numbers on the list less than 13, so 13 is our remainder. We add the numbers opposite the numbers we subtracted (16 + 4 + 1) to get our answer: 21. 1000 divided by 47 is 21 with a remainder of 13. That's Egyptian division.

PROBLEM 3.5   Use the Egyptian doubling method to divide 22,186 by 257.

PROBLEM 3.6   Use the Egyptian doubling method to divide 57,400 by 386.

Teachers and scribes used tables of previous results as a reference and shortcut, and some clever scribes undoubtedly discovered other shortcuts. Multiplying by 10 or a power of 10 just required substituting symbols. For 35×10, 3 heel bones (30) and 5 vertical strokes (5) are replaced by 3 coils of rope (300) and 5 heel bones (50). For 35×100, 3 heel bones (30) and 5 vertical strokes (5) are replaced by 3 lotus flowers (3000) and 5 coils of rope (500).

We don't know why the Egyptians were so fond of unit fractions, but they were, and the preserved records we have show that they went to great length to express other fractions in terms of unit fractions. Before we look at unit fractions, let's take a moment to describe the preserved records of Egyptian math. There are two kinds.

The first are mathematical references carved in stone. When Egyptian hieroglyphics were finally deciphered, the walls of tombs and temples and pyramids were *read*, and count of enemies defeated, years reigned, and wealth accumulated were part of the narrative.

Much more important for our understanding of how the Egyptians worked with numbers are the *papyri*. The plant *papyrus* was dried and pressed and converted into what we could call ancient paper (notice the similarity of the words *papyrus* and *paper*; and *papyri* is a better grammatical plural form than *papyruses*). Luckily for us, what was written on some of these remarkably well-preserved papyri was math. Dozens of math problems and their answers have been translated, and the papyri are now housed in museums around the world. You won't have much trouble guessing which museums own the Moscow papyrus or the Berlin papyrus; another spectacular one is the Rhind papyrus, named after the Scottish scholar Henry Rhind who bought it in the 19th century. Yes, it's in the British Museum now. We'll look at some papyrus problems later in this chapter. The papyrus problems were generally written with hieratic number symbols, which don't at all resemble the hieroglyphic symbols. Many students think the hieroglyphic symbols are cute or quaint. The hieratic symbols tend to look like messy ink blots. The Rhind papyrus is about 17-feet long and has arithmetic, algebra, and geometry problems. We even know the name of the scribe who copied it (Ahmes) and approximately when he copied it (1650 BC); but since it is a copy, we have to infer that the mathematics was well known somewhat before that date.

The first part of the Rhind papyrus has an elaborate list of unit fraction equivalents. For example, $\frac{2}{5}$ is shown to be $\frac{1}{3} + \frac{1}{15}$.

You can use a calculator and some guesswork to convert any fraction into a sum of unit fractions, but there are several useful techniques that are easy to learn and use.

The Egyptians apparently used several techniques, which are referred to now by memorably clever names, like "the greedy algorithm." In the greedy algorithm, the largest possible unit fraction is selected first. The unit fractions, from largest to smallest, are $\frac{1}{2}, \frac{1}{3}, \frac{1}{4}, \frac{1}{5}, \frac{1}{6}, \frac{1}{7}$, and so on. To convert $\frac{4}{7}$ to a sum of unit fractions, you would start with $\frac{1}{2}$ because $\frac{1}{2}$ is the largest unit fraction that is less than $\frac{4}{7}$.

$$\frac{4}{7} \quad \text{is} \quad \frac{1}{2} + \frac{1}{14}$$

To convert $\frac{3}{7}$ to a sum of unit fractions, you would start with $\frac{1}{3}$ because $\frac{1}{3}$ is the largest unit fraction less than $\frac{3}{7}$.

$$\frac{3}{7} \quad \text{is} \quad \frac{1}{3} + \frac{1}{14} + \frac{1}{42}$$

To convert $\frac{2}{7}$ to a sum of unit fractions, you would start with $\frac{1}{4}$ because $\frac{1}{4}$ is the largest unit fraction that is less than $\frac{2}{7}$.

$$\frac{2}{7} \quad \text{is} \quad \frac{1}{4} + \frac{1}{28}$$

Among the useful and easy ways to convert any fraction to the sum of unit fractions are the following three methods. Method 3.1 can be used when the denominator can be factored as a string of 2s and no other factors. (16, for example, is $2 \times 2 \times 2 \times 2$.) Method 3.2 can be used when the denominator has a sufficient variety of factors, and Method 3.3 can be used when the denominator lacks a sufficient variety of factors.

METHOD 3.1

This method works if the denominator is a power of 2. We add selected powers of $\frac{1}{2}$. The powers of $\frac{1}{2}$ are $\frac{1}{2}, \frac{1}{4}, \frac{1}{8}, \frac{1}{16}, \frac{1}{32}, \frac{1}{64}$, and so on.

You keep doubling the denominator to generate the next unit fraction in the series. Some successful conversions are as follows:

$$\frac{3}{4} = \frac{1}{2} + \frac{1}{4}$$

$$\frac{5}{8} = \frac{1}{2} + \frac{1}{8}$$

$$\frac{13}{16} = \frac{1}{2} + \frac{1}{4} + \frac{1}{16}$$

$$\frac{11}{32} = \frac{1}{4} + \frac{1}{16} + \frac{1}{32}$$

$$\frac{35}{64} = \frac{1}{2} + \frac{1}{32} + \frac{1}{64}$$

This method works because the numbers 1, 2, 4, 8, 16, 32, 64, and so on can be combined to equal any whole number. Consider the last fraction, $\frac{35}{64}$.

$$35 = 32 + 2 + 1.$$

So, $\frac{35}{64} = \frac{32}{64} + \frac{2}{64} + \frac{1}{64}$, and this becomes $\frac{1}{2} + \frac{1}{32} + \frac{1}{64}$

METHOD 3.2

This method works if the denominator has several factors and coincidentally the numerator is the sum of some of these factors. We express the denominator as various factor pairs, and look for sums of factors, hoping to find a sum that matches the numerator of the original fraction. It is often possible to find more than one way to express the original fraction.

Let's try $\frac{11}{18}$

$18 = \quad\quad 1 \times 18$

$\quad\quad$ or $\quad 2 \times 9$

$\quad\quad$ or $\quad 3 \times 6$

$11 = \quad (2+3+6), \text{ or } (2+9)$

So, $\frac{11}{18} = \frac{2}{18} + \frac{3}{18} + \frac{6}{18}$, or $\frac{2}{18} + \frac{9}{18}$

Which means $\frac{11}{18} = \frac{1}{9} + \frac{1}{6} + \frac{1}{3}$, or $\frac{11}{18} = \frac{1}{9} + \frac{1}{2}$

There is a pleasingly simple symmetry when we arrange the factors in a small table like this. The unit fractions we end up with have denominators that are opposite the numbers we add to make our sums.

We saw (for one answer) that 11 was the sum of 2, 3, and 6.

On the table of factors of 18, 2 is opposite 9, 3 is opposite 6, and 6 is opposite 3.

Our final answer was $\dfrac{1}{9}+\dfrac{1}{6}+\dfrac{1}{3}$.

Let's try $\dfrac{41}{50}$

$$50 = \quad 1 \times 50$$
$$\text{or} \quad 2 \times 25$$
$$\text{or} \quad 5 \times 10$$

$$41 = \quad 25 + 10 + 5 + 1$$

So, $\dfrac{41}{50}=\dfrac{25}{50}+\dfrac{10}{50}+\dfrac{5}{50}+\dfrac{1}{50}$

Which means $\dfrac{41}{50}=\dfrac{1}{2}+\dfrac{1}{5}+\dfrac{1}{10}+\dfrac{1}{50}$

METHOD 3.3

This method is essentially Method 3.2 with a preliminary step. We convert the original fraction to an equivalent fraction by multiplying its numerator and denominator by the same number, which doesn't change its value, and then use the same technique as Method 3.2. The purpose of multiplying the numerator and denominator is to increase the number of factors.

Observe that in Method 3.2, we cannot express $\dfrac{23}{50}$ as the sum of unit fractions, because no combination of 1, 2, 5, 10, and 25 equals 23.

But $\dfrac{23}{50}\times\dfrac{2}{2}=\dfrac{46}{100}$

$$\text{And } 100 = \quad 1 \times 100$$
$$\text{or} \quad 2 \times 50$$
$$\text{or} \quad 4 \times 25$$
$$\text{or} \quad 5 \times 20$$
$$\text{or} \quad 10 \times 10$$

$$46 = \quad 25 + 20 + 1$$

So, $\dfrac{46}{100} = \dfrac{25}{100} + \dfrac{20}{100} + \dfrac{1}{100}$

Or $\dfrac{23}{50} = \dfrac{1}{4} + \dfrac{1}{5} + \dfrac{1}{100}$

Another example is $\dfrac{7}{19}$. The denominator is a prime number. No factors at all.

Let's multiply by $\dfrac{6}{6}$. That creates lots of factors. $\dfrac{7}{19} \times \dfrac{6}{6} = \dfrac{42}{114}$

$$
\begin{array}{rrcr}
114 = & 1 & \times & 114 \\
\text{or} & 2 & \times & 57 \\
\text{or} & 3 & \times & 38 \\
\text{or} & 6 & \times & 19 \\
42 = & 38 & + \ 3 \ + & 1 \\
\end{array}
$$

So, $\dfrac{42}{114} = \dfrac{38}{114} + \dfrac{3}{114} + \dfrac{1}{114}$

Or $\dfrac{7}{19} = \dfrac{1}{3} + \dfrac{1}{38} + \dfrac{1}{114}$

Mathematicians have found many other ways to convert fractions to unit fractions, and in fact it has been proven that there are an unlimited number of possible conversions for any specific fraction. We saw before two possible conversions of $\dfrac{11}{18}$. Try these exercises that shows two more equivalents:

EXERCISE

Show that $\dfrac{11}{18} = \dfrac{1}{3} + \dfrac{1}{4} + \dfrac{1}{36}$

EXERCISE

Show that $\dfrac{11}{18} = \dfrac{1}{4} + \dfrac{1}{6} + \dfrac{1}{8} + \dfrac{1}{18} + \dfrac{1}{72}$

Suggestion: start by multiplying $\dfrac{11}{18}$ by $\dfrac{4}{4}$

PROBLEM 3.7　**Express** $\dfrac{19}{20}$ **as the sum of unit fractions.**

PROBLEM 3.8　**Express** $\dfrac{19}{24}$ **as the sum of unit fractions in two different ways.**

PROBLEM 3.9　**Express** $\dfrac{3}{17}$ **and** $\dfrac{15}{17}$ **as the sum of unit fractions.**

PROBLEM 3.10　**Express** $\dfrac{6}{13}$ **and** $\dfrac{10}{13}$ **as the sum of unit fractions.**

The Egyptian procedure for solving algebra problems involved making a guess and using a formula to "correct" the guess. This procedure is called *Egyptian false position*, and it goes like this:

$$\frac{1}{4}x - \frac{1}{6}x = 36$$

We want to solve for x, and we guess a value for x. Now, any guess will work, but it makes sense to make a guess that's easy to work with, and in this case we want a guess that eliminates the fractions, because whole numbers are easier to work with. So, a sensible guess here would be a number that is evenly divisible by 4 and by 6. Let's guess 24 and substitute it into the original equation and calculate

$$\left(\frac{1}{4}\cdot 24\right) \ - \ \left(\frac{1}{6}\cdot 24\right) \ = \ ??$$
$$6 \ \ - \ \ 4 \ \ = \ 2$$

We now have everything we need to find the real value of x. We have 36 (the original result), 24 (the guess), and 2 (the result when we used the guess). The formula is

$$\frac{\text{Original result}}{\text{Result from guess}}\left(\text{Guess}\right) = x$$

Substituting, we obtain

$$\frac{36}{2}\left(24\right) = 432$$

and you can verify that 432 solves the original equation.

EXERCISE

Try this problem with an original guess of 12 and with an original guess of 120.

MORE DIFFICULT EXERCISE

Prove algebraically that Egyptian false position works. Suggestion: start with the statements $\dfrac{a}{b}x = c$ **for the original equation, and** $\dfrac{a}{b}y = z$ **for the guess.**

Here, x is the unknown we are solving for.

$\frac{a}{b}$ **represents the (potentially complicated) rational expression in the original equation that we are multiplying x by, and c is the result of that operation (which is given in the original equation).**

In the second equation, y is the value that we guess for x, and z is the result of multiplying $\frac{a}{b}$ by y.

PROBLEM 3.11   Solve the following by false position:

$$\frac{1}{2}x+\frac{1}{3}x+\frac{1}{5}x=496$$

PROBLEM 3.12   Solve by false position:

$$\frac{1}{4}x+\frac{1}{5}x+\frac{1}{8}x=322$$

The papyri have many similar problems, some harder and some easier. Some problems are practical; others are clearly instructional, posing fanciful questions that were designed to sharpen students' skills.

One papyrus, called the Reisner papyrus, has practical problems. It looks like a business record for a building site. It lists employees and contains calculations for how many men are needed to do an excavation and what volume of dirt has to be removed.

The Rhind papyrus includes fanciful problems. In one, a man owns 7 houses and each house has 7 cats, and each cat catches 7 mice, and each mouse has 7 bunches of wheat, and each bunch has 7 grains. The student must total these things.

```
7   houses
7 × 7   =   49   cats
7 × 7 × 7   =   343   mice
7 × 7 × 7 × 7   =   2401   bunches
7 × 7 × 7 × 7 × 7   =   16,807   grains of wheat
```

Another simple problem from the Rhind papyrus can be solved easily by the false position technique: **a quantity and one fourth of the quantity equals 15; what is the quantity?**

We recognize this as $x+\frac{1}{4}x=15$

The modern way to do this problem is to say $\frac{5}{4}x=15$ and solve for x.

The Egyptian technique was to make a guess, and obviously guessing any multiple of 4 simplifies the problem by eliminating the fraction. If you guess 4, and substitute, you get $4+1=5$, and then the little formula gives you $\frac{15}{5}\cdot4=12$, which is the answer.

Another familiar problem asks us to **share 3 loaves of bread equally among 5 men**.

We would just say each man gets $\frac{3}{5}$ of a loaf of bread and be done with it, but the Egyptians didn't work so directly.

They would split one loaf into 5 parts, and give each man $\frac{1}{5}$.

They would split 2 loaves into thirds, making 6 thirds, and give each man $\frac{1}{3}$, accounting for 5 of the thirds. They would take that last third and split it into 5 parts, and give each man $\frac{1}{15}$.

So in the end, each man gets $\frac{1}{5} + \frac{1}{3} + \frac{1}{15}$.

This is one way to represent $\frac{3}{5}$ as the sum of unit fractions.

EXERCISE

Another way to represent $\frac{3}{5}$ is $\left(\frac{1}{2} + \frac{1}{10}\right)$. Show how 3 loaves can be cut up into halves and tenths and shared among the 5 men.

Here is a more complicated algebra problem from the Rhind papyrus: **a quantity plus $\frac{2}{3}$ of that quantity yields a sum; $\frac{1}{3}$ of that sum is subtracted, leaving 10; what is the quantity?**

$\frac{2}{3}$ is an interesting exception to the Egyptians' fixation on unit fractions. It is the only fractional quantity routinely expressed with a numerator other than 1. In modern terms, we might set this problem up as:

$$Q + \frac{2}{3}Q - \frac{1}{3}\left[Q + \frac{2}{3}Q\right] = 10$$

We can solve this equation by false position with a guess like 3 or 6 for Q, and we'll find that Q is 9. The equation simplifies to

$$\frac{2}{3}\left[Q + \frac{2}{3}Q\right] = 10$$

which we would solve by operating equally on both sides:

$$\left[Q + \frac{2}{3}Q\right] = 15 \quad \left(\text{multiplying both sides by } \frac{3}{2}\right)$$

$$\frac{5}{3}Q = 15 \quad \left(\text{combining terms}\right)$$

$$Q = 9 \quad \left(\text{multiplying both sides by } \frac{3}{5}\right)$$

A large number of problems concern mixtures and purity. Some of these look like practical problems. For example, **100 hekats of wheat with impurity 10 are to be exchanged for a volume of wheat of with impurity 45, and we are asked: what is a fair exchange?**

To begin, don't worry about *hekats*. It's a unit of measurement. You could think 100 pounds or 100 tons instead of 100 hekats. But we do have to interpret the phrases *impurity 10* and *impurity 45*. One straightforward interpretation is that impurity 10 means the 100 hekats include 90 hekats of pure wheat and 10 hekats of something else, probably something worthless, some filler or waste.

If we think a fair exchange of *wheat of impurity 10* and *wheat of impurity 45* means exchanging exactly equal amounts of wheat, then we need the volume of *wheat of impurity 45* that contains 90 hekats of wheat. 100 hekats of this inferior wheat contains 55 hekats of pure wheat, and 200 hekats of this inferior wheat contains 110 hekats of pure wheat, so the answer must lie somewhere between 100 and 200. We can in fact make a proportion and calculate the answer exactly: 163 and $\frac{7}{11}$ hekats.

But in the Rhind papyrus, this problem is interpreted differently. *Wheat with impurity 45* is said to be 4 and half times less valuable than *wheat with impurity 10* (since 45 is 4 and half times 10), and therefore four and a half times as much must be exchanged. The Rhind papyrus says 450 hekats is the answer, and we see this as a simple proportion:

$$\frac{10}{45} = \frac{100}{450}$$

Here is a rather challenging problem from the Rhind papyrus: **100 loaves are to be divided unequally among 5 men, and there is a common difference between each man's share. The two unluckiest (who get the smallest shares) together get only $\frac{1}{7}$ of what the three luckiest men get collectively. Find the shares.**

Before we tackle this, let's look at a very simple analogy. Let's say there were 15 loaves to be shared among 5 men, and the shares were 5, 4, 3, 2, 1. We can see there is a common difference of 1 loaf from share to share, and we can see the three top men get 12 loaves compared to the 3 loaves the bottom two men get, so the unluckiest two men get $\frac{1}{4}$ of what the three luckiest get in this scenario.

In a somewhat more complex scenario, the total loaves could be 55 and the shares could be 19, 15, 11, 7, 3. Here the common difference is 4 loaves from share to share, and the unluckiest two men get $\frac{2}{9}$ of what the three luckiest men get—they get 10 of the 55 loaves, while the other three men get 45 of the 55, and the ratio of 10 to 45 is $\frac{2}{9}$.

So, if you know the actual distribution, you can state the total, the common difference, and the ratio. In the papyrus problem, we only get the total (100 loaves), the fact that there is a common difference, and the ratio $\left(\dfrac{1}{7}\right)$.

The Egyptians solved this muddle using false position, but we can do better using modern algebra.

Let **s** = the share that the unluckiest guy gets

Let **d** = the difference between him and the next guy up

The five shares then are

$$\text{s}$$
$$\text{s} + \textbf{d}$$
$$\text{s} + 2\textbf{d}$$
$$\text{s} + 3\textbf{d}$$
$$\text{s} + 4\textbf{d}$$

The total is $5\text{s} + 10\textbf{d}$.
The conditions of the
problem tell us that $\quad 5\text{s} + 10\textbf{d} = 100 \quad \left[\text{equation for total}\right]$

$$\text{and that} \quad 2\text{s} + \textbf{d} = \frac{1}{7}(3\text{s} + 9\textbf{d}) \quad \left[\text{equation for ratio}\right]$$

We can exploit these relationships.

$$
\begin{array}{lll}
14\text{s} + 7\textbf{d} = 3\text{s} + 9\textbf{d} & \left(\text{multiplying the ratio equation by } 7\right) \\
14\text{s} \quad\quad\;\; = 3\text{s} + 2\textbf{d} & \left(\text{subtracting } 7\textbf{d} \text{ from both sides}\right) \\
11\text{s} \quad\quad\;\; = \quad\quad\; 2\textbf{d} & \left(\text{subtracting } 3\text{s} \text{ from both sides}\right) \\
5.5\text{s} \quad\quad\; = \quad\quad\;\; \textbf{d} & \left(\text{dividing by } 2, \text{ to get } \textbf{d} \text{ in terms of } \text{s}\right)
\end{array}
$$

$$
\begin{array}{lll}
5\text{s} + 10\textbf{d} \quad\quad = 100 & \left(\text{repeating the total equation}\right) \\
5\text{s} + 10(5.5\text{s}) = 100 & \left(\text{substituting for } \textbf{d}\right) \\
5\text{s} + 55\text{s} \quad\quad = 100 & \left(\text{distributing}\right)
\end{array}
$$

$$
\begin{array}{ll}
60\text{s} = 100 & \left(\text{combining terms}\right) \\[4pt]
\text{s} = \dfrac{100}{60} & \left(\text{dividing by } 60\right) \\[8pt]
\text{s} = \dfrac{10}{6} = \dfrac{5}{3} & \left(\text{reducing the fraction}\right)
\end{array}
$$

Since we know **s**, we can find **d**:

$$\mathbf{d} = 5.5s = (5.5)\left(\frac{5}{3}\right) = \frac{27.5}{3} = \frac{55}{6} \quad \left(\text{this is just over } 9\right)$$

So the shares are: $\frac{5}{3}$, and multiples of $\frac{55}{6}$ added to $\frac{5}{3}$.

From top to bottom, this is

$\dfrac{115}{3}$	38 and $\dfrac{1}{3}$
$\dfrac{175}{6}$	29 and $\dfrac{1}{6}$
20	(exactly 20)
$\dfrac{65}{6}$	10 and $\dfrac{5}{6}$
$\dfrac{5}{3}$	1 and $\dfrac{2}{3}$

The Egyptians of course would have found a different way to express the fraction $\frac{5}{6}$ in the fourth man's share, using unit fractions. His share is $10 + \frac{1}{2} + \frac{1}{3}$.

PROBLEM 3.13   **150 loaves of bread are to be divided *unequally* among 6 men, with a *fixed difference* between consecutive shares. The least fortunate man gets only $\frac{1}{4}$ as many loaves as the most fortunate man.**

What are the 6 shares?

PROBLEM 3.14   **210 loaves of bread are to be divided *unequally* among 5 men, with a *fixed difference* between consecutive shares. The two least fortunate men together get only $\frac{1}{6}$ as many loaves as three most fortunate men (together).**

What are the 5 shares?

From the Berlin papyrus, we have this interesting challenge: to divide a square of area 100 into two new smaller squares so that the side of the smaller new square is $\frac{3}{4}$ the length of the side of the larger new square.

The Egyptians used false position to find that the new squares had sides of 6 and 8.

In modern terms, we might express this as $6^2 + 8^2 = 10^2$. This suggests the Egyptians were familiar with some Pythagorean right triangles. Of course, a square cannot be divided into two smaller squares without rearranging some fragments, but the author of the Berlin papyrus must have played with the shapes and visualized something like this:

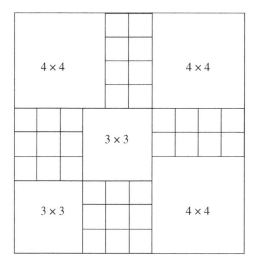

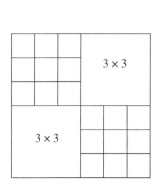

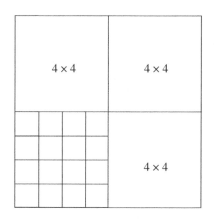

FIGURE 3.4   Dissecting the square in the problem from the Berlin papyrus.

The large square at the top of Figure 3.4 has an area of 100. It is composed of many smaller squares, but you can see that it is basically a (10×10) grid of small squares. Some of the 100 small squares have been collected and organized to make medium-sized squares: there are (4×4) squares in three of the corners (northwest, northeast, and southeast), and there are (3×3) squares in the middle and in the southwest corner. There are four groups of small squares that have not been collected and organized. The two groups that look like little tic-tac-toe boards are themselves (3×3) squares, and the other two groups are each equivalent to half of a (4×4) square.

You can visualize breaking the square at the top of Figure 3.4 into the two squares at the bottom of Figure 3.4 by first pulling all the (3×3) shapes away to the left. The remaining (4×4) squares can be squeezed together, with the 16 displaced little squares reassembled to make another (4×4) square. The resulting shapes satisfy the requirement in the problem from the Berlin papyrus. On the left side, we see a (6×6) square composed of four (3×3) squares; on the right, we see an (8×8) square composed of four (4×4) squares. Every (3×3) square on the left has a corresponding (4×4) square on the right we can match it to, so we have divided the original square of area 100 into two squares with the proper ratio.

We know that the Egyptians could calculate the volume of part of a pyramid because there is a problem in the Moscow papyrus where such a calculation is done. If you slice off the top of a square pyramid, the part that is left is called a *frustum*. It has a square base and a square top and four trapezoidal sides.

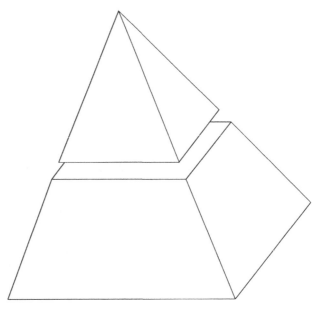

FIGURE 3.5   Slicing off the top of a pyramid creates a frustum.

If you are in the business of building big pyramids (and the Egyptians were in that business), then you tend to have a lot of frustums around while your work is in progress. And sometimes, you want to know the volume of one of the frustums. If the side of the top square is **a** and the side of the bottom square is **b**, and the height of the frustum is **h**, then the formula for the volume of the frustum is

$$\frac{1}{3}\mathbf{h}\left(\mathbf{a}^2+\mathbf{ab}+\mathbf{b}^2\right).$$

The frustum shown in the Moscow papyrus has these dimensions: the square at the top has a side of length 2, the square at the base has a side of length 4, and the height of the frustum is 6. The calculation for the volume of this frustum is

$$\frac{1}{3}(6)\left(2^2+2\cdot4+4^2\right),\ \text{which is }56.$$

EXERCISE

Confirm, using similar triangles, that if this frustum was extended upward to make a complete pyramid, the height of this pyramid would be 12. Using the Greek formula for the volume of a pyramid, which is $V = \dfrac{1}{3} B \cdot h$, where B is the area of the base, find the volume of this complete pyramid. Again using the Greek formula, confirm that the volume of the frustum is 56 by subtracting the little pyramid added on top of the frustum to make the complete pyramid.

Just as we have our superstitions about numbers (7 is thought to be lucky, and 13 is thought to be unlucky), the Egyptians had their preferences. For them, 28 was a wonderful number and 17 a bad one. One very appealing property of 28 is that if you add up the numbers that divide evenly into 28, you get 28. This is a rare mathematical property. $28 = 1 + 2 + 4 + 7 + 14$, and those five numbers are the only ones that go into 28 evenly. The Greeks, who studied relationships like this in much more depth, called 28 a *perfect* number because of this coincidence. Undoubtedly, the Egyptians also knew that 6 was a perfect number in this sense and therefore liked 6. $(6 = 1 + 2 + 3)$

28 is also the number of days the moon takes to renew itself, going from full moon to new moon to full again, through reliably predictable stages. The Egyptians, like virtually every other ancient society, had some reverence for the moon and regarded it as godlike. If you can visualize what the world was like before man-made lighting (electricity, and before that burning gases), you can appreciate what a wonder the moon was. Ancient people thought it generated light on its own, while we of course know now that moonlight is actually reflected sunlight. The number 28 figures in Egyptian mythology as well.

The dislike of 17 had a mythological basis too, but 17 is importantly unlike its neighbors among the counting numbers, 16 and 18. 16 can be expressed as 2×8 or as 4×4; and 18 can be expressed as 2×9 or as 3×6. Thus, 16 things could be split evenly among 2 or 4 people, and 18 things could be shared equally among 2, 3, or 6 people. In modern terms, we call 17 a prime number. 17 things cannot be divided into equal shares, without one person getting more than the others. 17 was a bad interloper in between two very good numbers.

There is a further geometrical reason for the Egyptians liking 16 and 18: these are the only two numbers that can simultaneously be the perimeter and the area of a rectangle.

One Egyptian geometry puzzle poses exactly that question: find the dimensions of a rectangle whose perimeter (in units of length) matches its area (in square units). The Egyptians found the only two rectangles that satisfy this challenge (using whole numbers only), but as far as we know they didn't have any proof that more could not be found. In modern terms, we can call the sides of the rectangle **a** and **b**. The perimeter therefore is **2a + 2b**, and the area is **ab**.

So, finding the dimensions of the rectangles means solving the equation **2a + 2b = ab**.

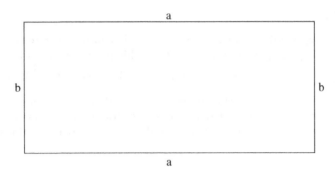

FIGURE 3.6   A rectangle with length **a** and width **b**. The Egyptians sought values for **a** and **b** that would make the perimeter and area of the rectangle equal.

Let's do something the Egyptians never did (but the Greeks did all the time): operate on both sides of the equation. Subtracting **2b** from both sides gives

$$2a = ab - 2b$$

This can be factored a bit on the right side:

$$2a = b(a - 2)$$

And if we divide both sides by $(a - 2)$, we get this expression for **b**:

$$b = \frac{2a}{(a - 2)}$$

We could just as easily express **a** in terms of **b**, but the analysis is the same whichever variable we focus on. Let's think about what the equation above tells us about **b**.

b has to be a whole number because that's a condition the Egyptians wanted, and **a** also has to be a whole number. Further, **a** has to be greater than 2; otherwise the denominator of the fraction would be an unknown number (the Egyptians did not know about zero or negative numbers). So, let's experiment with the possible values for **a**:

If **a** = 3, the fraction becomes $\frac{6}{1}$, and we have the solution **a** = 3 and **b** = 6.

If **a** = 4, the fraction becomes $\frac{8}{2}$, and we have the solution **a** = 4 and **b** = 4.

If **a** = 5, the fraction becomes $\frac{10}{3}$, and we have a bad value for **b**.

If **a** = 6, the fraction becomes $\frac{12}{4}$, and we have the solution **a** = 6 and **b** = 3.

If **a** = 7, the fraction becomes $\frac{14}{5}$, and we have a bad value for **b**.

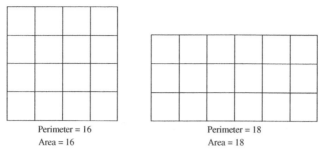

Perimeter = 16   Perimeter = 18
Area = 16   Area = 18

FIGURE 3.7 The 4×4 square and the 3×6 rectangle.

As **a** gets larger and larger, **b** gets closer and closer to 2, without ever reaching exactly 2, so there are no more pairs of **a** and **b** that solve the equation with whole numbers.

The solutions, popular among ancient Egyptian mathematicians, are a square (4×4) and a rectangle (3×6) that looks like two squares (3×3) joined side to side.

PROBLEM 3.15 **Let's make a variant of the Egyptian requirement. Find the dimensions of a rectangle whose area (in square units) is two times its perimeter (in units of length). Find three distinct solutions.**

PROBLEM 3.16 **Find the dimensions of a rectangle whose area (in square units) is three times its perimeter (in units of length). Find five distinct solutions.**

Suggestions for book or Internet research	
History, archeology	Ancient Egypt
	River valley civilizations
	Rosetta stone
	Papyrus
Religion, culture	Isis and Osiris
Mathematical topics	Egyptian mathematics
	Egyptian hieroglyphic numerals
	Egyptian multiplication
	Egyptian fractions
	Rhind papyrus
	Moscow Papyrus, Berlin Papyrus

ANSWERS TO PROBLEMS

3.1

Start with the numbers 1 and 67 This represents that $1 \times 67 = 67$	1	67
Double both numbers and write the results underneath. This represents that $2 \times 67 = 134$	1 2	67 134

Continue to double both columns until	1 67
the number in the first column is	2 134
greater than 42 (because the problem	4 268
is 67×42)	8 536
	16 1072
	32 2144
	64 4288

Find numbers in the first column that add up to 42.

$$32 + 8 + 2 = 42.$$

Add the numbers from column two that are opposite 32, 8, and 2.
$2144 + 536 + 134 = 2814$. This is the correct answer.
It is the sum of thirty-two **67s**, and eight **67s**, and two **67s**.

3.3

Start with the numbers 68 and 43.	68 43
Cut 68 in half and double 43	34 86
Cut 34 in half and double 86	17 172
Cut, and double	8 344
Notice that half of 17 is $8\frac{1}{2}$, but we ignore the fraction	
Continue to cut and double, until the first column is down to 1.	4 688
	2 1376
	1 2752

At this point, look for odd numbers in column 1. The only odd numbers are 17 and 1.
So we add the numbers in column 2 that are opposite 17 and 1.
$172 + 2752 = 2924$. This is the correct answer.

3.5

22,186 divided by 257. Start by doubling 1 and 257 until the number in the second
column is larger than 22,186.

1	257
2	514
4	1028
8	2056
16	4112
32	8224
64	16448
128	32986

Subtract the biggest number in column 2 that is less than 22,186 from 22,186

$$\begin{array}{r} 22186 \\ \underline{16448} \\ 5738 \end{array}$$

Since 16,448 was opposite 64, 64 is part of our answer. Now we subtract the biggest number in column 2 that is less than 5738

$$5738$$
$$\underline{4112}$$
$$1626$$

Since 4112 was opposite 16, 16 is part of our answer. Now we subtract the biggest number in column 2 that is less than 1626

$$1626$$
$$\underline{1028}$$
$$598$$

Since 1028 was opposite 4, 4 is part of our answer. Now we subtract the biggest number in column 2 that is less than 598

$$598$$
$$\underline{514}$$
$$84$$

Since 514 was opposite 2, 2 is part of our answer. There are no numbers in column 2 that are less than 84, so the series of subtractions is complete and 84 is a remainder.

We add $64 + 16 + 4 + 2$ to get our answer: 22,186 divided by 257 is 86 with a remainder of 84.

3.7

$\dfrac{19}{20}$. Start by listing the factors of 20

1	20
2	10
4	5

We need a combination of the factors of the denominator to add up to the numerator. Here, $4 + 5 + 10 = 19$.

So, $\dfrac{4}{20} + \dfrac{5}{20} + \dfrac{10}{20} = \dfrac{19}{20}$

Or $\dfrac{19}{20} = \dfrac{1}{5} + \dfrac{1}{4} + \dfrac{1}{2}$

3.9

$\dfrac{3}{17}$ and $\dfrac{15}{17}$

17 is a prime number, so its only factors are 1 and 17. Obviously, no combination of 1 and 17 adds up to 3 or 15. So, we multiply these fractions by 1 in the form of $\frac{6}{6}$ to make equivalent fractions. Multiplying by $\frac{6}{6}$ often works.

$$\frac{3}{17} \times \frac{6}{6} = \frac{18}{102}$$

$$\frac{15}{17} \times \frac{6}{6} = \frac{90}{102}$$

The new denominator is 102, which has many factors. The factors of 102 are

1	102
2	51
3	34
6	17

Once again, to make unit fractions, *we need a combination of the factors of the denominator to add up to the numerators.* The numerators are 18 and 90.

$$18 = 1 + 17, \quad \text{so} \quad \frac{18}{102} = \frac{1}{102} + \frac{17}{102},$$

$$\text{or} \quad \frac{3}{17} = \frac{1}{102} + \frac{1}{6}$$

$$90 = 2 + 3 + 34 + 51, \quad \text{so} \quad \frac{90}{102} = \frac{2}{102} + \frac{3}{102} + \frac{34}{102} + \frac{51}{102},$$

$$\text{or} \quad \frac{15}{17} = \frac{1}{51} + \frac{1}{34} + \frac{1}{3} + \frac{1}{2}$$

Multiplying by $\frac{6}{6}$ worked in this case. It won't always work. When it doesn't work, try something else, like $\frac{12}{12}, \frac{18}{18}$, or $\frac{20}{20}$. The goal is to get a large enough pool of factors that some combination of them matches the numerator.

EXERCISE

Verify that $\frac{8}{17}$ cannot be expressed as the sum of unit fractions when you multiply it by $\frac{6}{6}$, but can be expressed as the sum of unit fractions when you multiply it by $\frac{20}{20}$.

3.11

Any multiple of 30 is a sensible guess, because the denominators of the unit fractions in the equation multiply to 30. (Since any guess will work, make a guess that's easy to work with.)

$$\text{Plugging } 30 \text{ in for } x, \text{we get:} \quad \left(\frac{1}{2} \cdot 30\right) \ + \ \left(\frac{1}{3} \cdot 30\right) \ + \ \left(\frac{1}{5} \cdot 30\right)$$

$$= \quad 15 \quad + \quad 10 \quad + \quad 6$$

$$= \quad 31$$

So, we take the original result, divided by 31, and multiplied by the guess (30):

$$\frac{496}{31} \times 30 = 480, \text{ and } 480 \text{ is our answer.}$$

3.13

The technique is exactly the same as the technique shown in the chapter. If s represents the share of the least fortunate man, and d represents the consecutive difference between shares, the key equations are

$$6s \ + \ 15d \ = \ 150$$

$$s \quad\quad\quad = \quad \frac{1}{4}\left(s + 5d\right)$$

Solving this system of equations yields the result that

$$s = 10, \text{ and } d = 6$$

so the men's shares are

$$10$$
$$16$$
$$22$$
$$28$$
$$34$$
$$40$$

3.15

We need to solve the equation 2 [perimeter] = [area]
Symbolically, that's $2\left[2a + 2b\right] = ab$

$$\text{Or, } 4a + 4b = ab$$

Subtracting **4b** from both sides gives

$$\mathbf{4a = ab - 4b}$$

This can be factored a bit on the right side:

$$\mathbf{4a = b\left(a - 4\right)}$$

And if we divide both sides by $\left(\mathbf{a - 4}\right)$, we obtain the following expression for **b**:

$$\mathbf{b = \frac{4a}{\left(a - 4\right)}}$$

b has to be a whole number, and **a** also has to be a whole number. Further, **a** has to be greater than 4. Experimenting with the possible values for **a** we find the following:

If **a** = 5, the fraction becomes $\dfrac{20}{1}$, and we have the solution **a** = 5 and **b** = 20.

If **a** = 6, the fraction becomes $\dfrac{24}{2}$, and we have the solution **a** = 6 and **b** = 12.

If **a** = 8, the fraction becomes $\dfrac{32}{4}$, and we have the solution **a** = 8 and **b** = 8.

The 5×20 rectangle has a perimeter of 50 and an area of 100.

The 6×12 rectangle has a perimeter of 36 and an area of 72.

The 8×8 square has a perimeter of 32 and an area of 64.

4

ANCIENT CHINESE MATHEMATICS

The development of math in ancient China is poorly documented. Partly this is because the oldest records were not preserved well. Whereas the Babylonians left us thousands of hardened clay tablets and the Egyptians left us sturdy papyrus, the ancient Chinese relied on flimsier paper and bamboo. Think about what you would know about your own life and ancestry if you relied only on paper records. You might have trouble just finding paper records of your own early childhood, and if you searched for paper records of your parents' and grandparents' lives, you would find that the vast majority of the details have not been preserved. So, imagine the difficulties you would have if you were looking for paper records going back thousands of years.

The other reason for poor documentation has to do with imperial Chinese culture. Old records were deliberately destroyed. Many emperors, when ascending to power, believed that their own glories would be enhanced by removing evidence of previous accomplishments, under previous rulers. This led to burning the books that previous administrations used, and these books might have included profound and vital guides to law, commerce, engineering, government, history, and so on, and there is no reason to think books about the sciences (astronomy and mathematics) would be spared. What we see again and again in the surviving books are references to earlier books, now lost. We infer the mathematical results and accomplishments of previous generations by observing how one mathematical writer explains that his results are an improvement on an earlier writer's commentary on yet an even earlier one's technique.

An Introduction to the Early Development of Mathematics, First Edition. Michael K. J. Goodman.
© 2016 John Wiley & Sons, Inc. Published 2016 by John Wiley & Sons, Inc.

Among the few books that we know by title and content is *Nine Chapters on the Mathematical Art*. It contains (unsurprisingly) nine chapters: the first chapter is mostly about surveying, and a typical problem is to calculate the area of a rectangular field. The second chapter is mostly about rice, and a typical problem is to calculate a fair price. The nine chapters contain over 200 problems that cover practical and theoretical aspects of algebra and geometry. It appears to have been written between 2000 and 3000 years ago, with the possibility that its oldest portions were written centuries before other scholars added to it. The version we know was put together by the mathematician Liu Hui in AD 263.

Among the subjects Chinese mathematicians wrote about are fractions, proportions, square roots and cube roots, volumes of trapezoidal mud pits, tax rates, right triangles, the flow of water, probability, magic squares, and systems of equations. The sophistication of some of the ideas suggests that they were continuations of a long tradition of mathematical study. This seems pretty reasonable. You wouldn't expect expertise about adding fractions unless you already had expertise in adding and expertise in fractions. So, where we see sophistication, we infer preliminary work at a more basic level. What is unclear is how much time it took for the Chinese to progress from the basics to the higher levels.

In addition to the books, there is another curious source of material about Chinese math: tortoise shells. A *tortoise* is a turtle that lives chiefly on land, and the front part of its shell is called a *plastron* and the back part is called a *carapace*. Archeologists made tremendous progress in understanding ancient Chinese math when, in 1899, they discovered thousands of plastrons with legible inscriptions from the Shang dynasty. Apparently it is pretty easy to write on a plastron. It doesn't seem too easy to write on a carapace, but there is a legend that one emperor was trapped by enemies and made his escape when a turtle ferried him across a river, a turtle that happened to have a 3×3 magic square etched into its shell, which may be the justification for the enthusiasm Chinese mathematicians had for creating magic squares. The properties of magic squares were investigated and well understood, and Chinese students mastered various techniques for constructing them.

A fair historical question is whether the Chinese accomplished what they accomplished before or after other civilizations did, and whether they did it in isolation or were influenced by other cultures. The lack of documents obviously limits our ability to answer this. But it seems that goods were exchanged for thousands of years via that network of trading routes across central Asia collectively known as the Silk Road, and just as goods made their way east and west, ideas similarly traveled. Certainly the math needed to do buying and selling between merchants with different coinages would quickly diffuse, as would methods of calculating distance and time. The safest answer is that some exchange of intellectual property occurred, but some things were discovered again and again by different people at different times.

In the early 1980s a 1000-year-old tomb was excavated in China, and that tomb contained 200 bamboo strips with arithmetic problems. Some of the problems are like the ancient Egyptian "sharing" problems where things are split up and everyone's share in the end is the sum of some unit fractions. It is conceivable that some arithmetical ideas went from Egypt to China with a hundred stops along the way, but it seems more likely that every society that developed counting and division eventually thought about how to express fractions and how to teach fractions and came up with the stereotypical "sharing" problem with unit fractions as a student exercise.

Let's turn briefly to the Chinese number systems, which to most westerners are an obstacle. Two main systems are known. The first, a calligraphy system, intimidates westerners because of its unfamiliar symbols. The second, a straight line system, looks too crude to be very useful. But both are easily capable of expressing the very large numbers the Chinese needed to run their empire. Large numbers were required simply because of the size of the country and the number of people and the scale of civic projects. Chinese bureaucrats had to deal accurately with massive harvests, taxes, and military enterprises. The practical needs of administering a city or an empire spurred mathematical thought.

The traditional Chinese numerals we study today were standardized during the Han dynasty, perhaps in 200 BC. They differ somewhat from Shang dynasty numerals (perhaps from 1100 BC) but are clearly related to and derived from them. The key principle is that two characters together can make what we would call a digit. Where we say 653, they have a symbol for 6 and a symbol for hundred, to tell us 600, and they have a symbol for 5 and a symbol for 10, to tell us 50, and they have a symbol for 3 (which doesn't need a companion symbol since the 3 is not being multiplied by a power of 10). Our 3-digit number is expressed in 5 Chinese calligraphy characters.

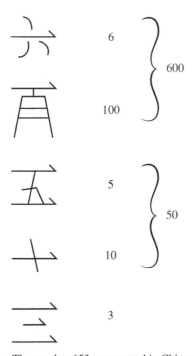

FIGURE 4.1 The number 653 represented in Chinese calligraphy.

Now, remarkably, this system can be used by western students after only a little practice, and it is recommended that the student imagine himself or herself as an apprentice scribe in an ancient Chinese school. Practicing making the symbols, and practicing making simple arithmetical computations with the symbols, will give the student an appreciation of both the usefulness of the system and the hard work required to be a successful scribe.

Only simple computations need be made with the symbols, as the Chinese used devices like counting boards to work out difficult computations. Counting boards are somewhat like the calligraphy symbols themselves: strange at first sight, but useful and natural when one gets used to them.

The counting boards were flat, rectangular wooden arrays, and they were split into compartments that represented the columns of numbers we learn in elementary school (the ones column, the tens column, the hundreds column, etc.). The compartments could be filled with stones or other counters or markers, and the scribe could gather and shift the stones to reflect mathematical operations. Let us imagine that the scribe wanted to add 536, 178, and 429. He could start by placing the stones like this:

		OOOOO	OOO	OOOOOO
		O	OOOOO OO	OOOO OOOO
		OOOO	OO	OOOOOO OOO

FIGURE 4.2 The start of the addition problem $536 + 178 + 429$.

Observe that at this point the three rows of compartments represent the three numbers to be added, with the units digits on the extreme right and the other digits occupying columns further to the left. It doesn't matter at the point how the scribe groups the stones within each compartment.

Next, the scribe moves stones, collecting stones that are in the same column. He might choose to move all the stones to the bottom row.

		OOOOO O OOOO	OOOOO OOOOO OO	OOOOOO OOOOOO OOOOOO OOOOO

FIGURE 4.3 The stones have all been moved to the bottom row.

Because the columns represent 10,000s, 1,000s, 100s, 10s, and 1s, ten stones in one column can be replaced by one stone in the column immediately to its left. So an acceptable transfer is:

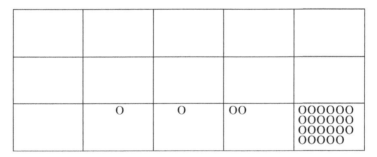

FIGURE 4.4   Ten stones in the hundreds column have been replaced by a single stone in the thousands column.

Ten hundreds have been replaced by one thousand.
There are more than 10 stones in the tens column, so another acceptable transfer is:

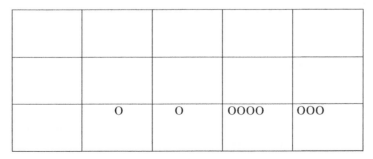

FIGURE 4.5   Ten stones in the tens column have been replaced by a single stone in the hundreds column.

And two groups of ten stones (in the ones column) can be replaced by two stones in the tens column:

FIGURE 4.6   Twenty stones in the ones column have been replaced by two stones in the tens column.

This manipulation would tell our scribe that 536, 178, and 429 add up to 1143.

Notice that we started the addition work in the hundreds column (carrying to the thousands), and worked our way eventually to the ones column. That is the Chinese way. American schoolchildren are taught to work in the opposite direction, from the ones column upward to the hundreds column. Chinese mathematicians worked with the high-order digits first so they would quickly get a good estimate of the final answer. If we were to add 7846+5619 the way American schoolchildren are taught to, our first operation would be to add the 6 and the 9. This pinpoints the least significant digit but does not say anything about the magnitude of the sum. The Chinese approach, adding 7000+5000 first, tells them immediately approximately how big the sum will be.

Counting boards were put to much more sophisticated uses than what we just saw: they were used to do multiplication problems and to find square roots. In multiplication, the same principle of computing high-order digits first was followed. If we today were to multiply 7846×5619 (by hand), the very first calculation we would make is 9×6, and we might think, "54, write the 4, carry the 5…" and we'd have no idea what our approximate final result would be until we had carried out many more steps. In the old Chinese method, the very first step is 7000×5000, and right away the student knows the answer is well over 35 million. In fact 7846×5619 is over 44 million.

Here are the traditional Chinese calligraphy numbers.

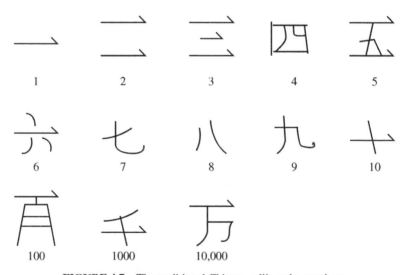

FIGURE 4.7   The traditional Chinese calligraphy numbers.

It looks like some of these drawings are random, but there are elements that occur multiple times. The most obvious recurring element is the horizontal arrow, which logically appears one, two, and three times in the designs for the numbers 1, 2, and 3. But the horizontal arrow also is part of 5, 6, 10, 100, 1,000, and 10,000.

The box shape around 4 is similar to the box shape around 100. Within the box (in the symbol for 4), there is a J-shape and there is an L-shape. The J-shape is seen again in the 9 and the 10,000. The L-shape is seen again in the 7 and the 9.

There is a curving line common to the 6 and the 8, and a vertical line common to the 10 and the 1000. The 5 and 9 share a short horizontal line connected to an L-shape.

You will observe that generally in the calligraphy system there is no need for a zero. In making the characters for a number like 5704, you would write the symbols for 5 and 1000, and 7 and 100, and 4. Putting in symbols for 0 and 10 would be superfluous.

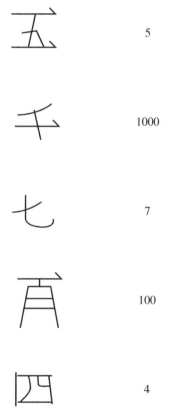

	5
	1000
	7
	100
	4

FIGURE 4.8 The number 5704 represented in Chinese calligraphy.

But eventually several symbols for zero were added to the system. The Chinese symbols, like the Egyptian symbols, evolved over time. Perhaps, a stimulus to represent zero directly came from the empty spaces that occurred on counting boards.

Let's work through the addition problem represented in Figure 4.9.

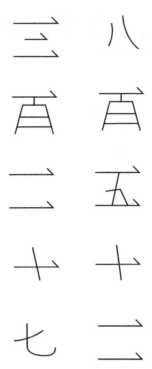

FIGURE 4.9   The addition problem 327 + 852.

The first column represents 327. The second column represents 852. Chinese numbers were typically written vertically, with the largest digit on top.

The first thing a Chinese scribe would do is add the hundreds. He would see he has 3 hundreds in the first column and 8 hundreds in the second column, which together make 11 hundreds. He would know to write this as 1 thousand and 1 hundred. Notice that he would only need to put in the symbols for a thousand and a hundred. Technically, it wouldn't be wrong to put the symbol for 1 over the symbol for 1000 and the symbol for 1 over the symbol for 100, but it is unnecessary to do so. The bracket in Figure 4.10 is not part of the calligraphy—it is just a convenience to highlight which part of the problem we are working on.

FIGURE 4.10 300 is added to 800.

Next our scribe would add the tens, showing that 2 tens and 5 tens make 7 tens.

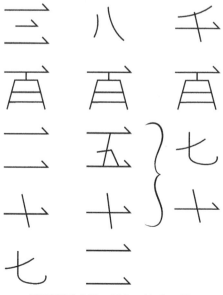

FIGURE 4.11 20 is added to 50.

Finally, the scribe would add 7 and 2 to get 9. The final result is 1179.

FIGURE 4.12   7 is added to 2 to complete the problem.

Here is an addition problem that involves more steps, because we have to "carry" or convert symbols:

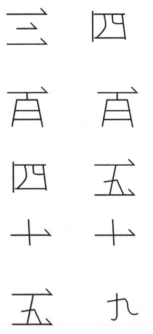

FIGURE 4.13   The addition problem 345 + 459.

The problem is 345 + 459. Again, the first thing to do is add the hundreds. We see that 3 hundreds and 4 hundreds make 7 hundreds. Then we see that 4 tens and 5 tens make 9 tens. Then we see that 5 and 9 make 14, which is written as 10 over 4.

FIGURE 4.14   300 and 400 have been added; 40 and 50 have been added; and 5 and 9 have been added.

Something is fishy in Figure 4.14 because, reading down, we see "seven hundreds, nine tens, ten and four." We should never see the same symbol repeated, and here there are two symbols for 10 next to each other. We combine them. 9 tens and 1 ten make 10 tens, which is 100.

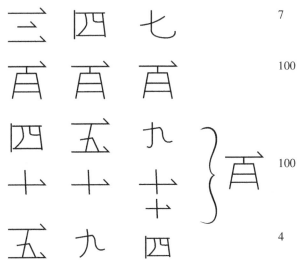

FIGURE 4.15   90 and 10 are combined.

Things are still not quite right. Reading down in Figure 4.15, we see "seven hundreds, a hundred and four." Again we have two identical symbols next to each other: the symbol for 100 is repeated, so we combine 700 and 100.

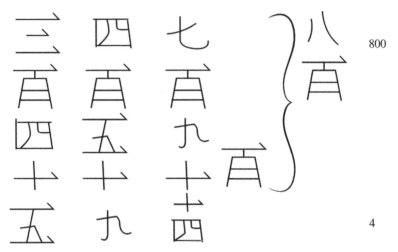

FIGURE 4.16   700 is added to 100, completing the problem.

Our final answer has only 3 symbols, even though we started with two numbers that each required 5 symbols. You can verify that we just computed 345 + 459 = 804.

The second well-known Chinese number system is called variously Chinese Scientific, Chinese Bamboo, or Chinese Rod, and its key feature is its simplicity: the symbols do not require elaborate drawing and memorization. Here is a large number expressed with the straight-line logic of the bamboo rods:

FIGURE 4.17   The number 867,308 represented in Chinese scientific symbols.

This number is 867,308. A 6-digit number in our system is represented by 5 characters here. The leftmost character means 800,000. Next comes a T-shape for 60,000. Next comes a symbol for 7000. Then 3 vertical lines for 300. Last, a symbol for 8.

Notice that the 8 that represents 800,000 and the 8 that represents 8 are similar but not exactly the same. In the 800,000 symbol, the 3 parallel lines are horizontal; in the 8 symbol, the 3 parallel lines are vertical. In the Chinese Scientific system, the characters representing 1s, 100s, 10,000s, and other "odd" positions have a vertical orientation; the characters representing 10s, 1,000s, 100,000s, and other "even" positions have a horizontal orientation. There is a branching or T-shaped element for digits greater than 5.

In our 867,308 there is a zero in the tens position, and in their 867,308 the last two characters are oriented vertically, indicating that a horizontally oriented symbol has been omitted, their way of implying that a zero has been left out.

The scheme for the symbols is very straightforward. Here are the numbers that represent our 1 through 9:

$$| \quad || \quad ||| \quad |||| \quad ||||| \quad \top \quad \overline{\top} \quad \overline{\overline{\top}} \quad \overline{\overline{\overline{\top}}}$$

$$\begin{array}{ccccccccc} 1 & 2 & 3 & 4 & 5 & 6 & 7 & 8 & 9 \end{array}$$

FIGURE 4.18   The Chinese scientific symbols for ones, hundreds, and ten thousands.

These exact symbols also represent 100, 200, 300, 400, 500, 600, 700, 800, and 900.

They also represent the multiples of 10,000: 20,000, 30,000, and so on up to 90,000.

Here are the numbers that represent 10, 20, 30, 40, 50, 60, 70, 80, and 90:

$$\underline{\quad} \quad \underline{\underline{\quad}} \quad \underline{\underline{\underline{\quad}}} \quad \underline{\underline{\underline{\underline{\quad}}}} \quad \underline{\underline{\underline{\underline{\underline{\quad}}}}} \quad \perp \quad \underline{\!\!\perp\!\!} \quad \underline{\!\!\!\perp\!\!\!} \quad \underline{\!\!\!\!\perp\!\!\!\!}$$

$$\begin{array}{ccccccccc} 10 & 20 & 30 & 40 & 50 & 60 & 70 & 80 & 90 \end{array}$$

FIGURE 4.19   The Chinese scientific symbols for tens, thousands, and hundred thousands.

These exact symbols also represent 1000, 2000, 3000, and so on up to 9000.

And 100,000, 200,000, 300,000, and so on up to 900,000.

Since so many symbols are identical, their position matters a great deal. And the absence of a zero symbol makes some numbers a little ambiguous. In our modern system, we distinguish between 404 and 40,004 by the number of zeroes between the fours. In Chinese Scientific, it's only the amount of space between the fours that makes them different.

404   40,004

FIGURE 4.20   The numbers 404 and 40,004 represented in Chinese scientific symbols.

There are many clever problems and theorems among the preserved Chinese material. One of the most delightful geometrical discoveries concerns the equivalence of rectangles. Consider Figure 4.21, in which a large rectangle is divided into smaller rectangles by intersecting lines. The question arises: are the areas of rectangles A and B equal?

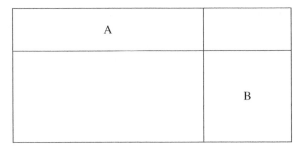

FIGURE 4.21   Are the smaller rectangles, A and B, equal in area?

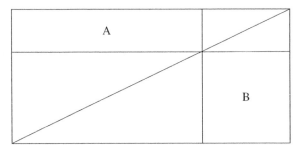

FIGURE 4.22   The Chinese proof that the areas are equal.

The Chinese found a wonderfully simple answer to this question. The horizontal and vertical lines that divide the original rectangle meet at a point. If that point lies on the rectangle's diagonal, then A and B are equal in area; otherwise they are not. The proof is in Figure 4.22.

Diagonals split rectangles into equal halves. If the point where the horizontal and vertical lines meet is on the diagonal of the large rectangle, it is also on the diagonals of the smaller unmarked rectangles and also splits them into equal halves. Above the diagonal are rectangle A and two triangles, and below the diagonal are rectangle B and two triangles, and these triangles, in pairs, are equal parts of the unmarked rectangles. The diagonal splits the large rectangles into equal (triangular) halves, and subtracting equals from equals leaves rectangle A and rectangle B, which therefore must be equal.

Let's look at one technique for making magic squares. Figure 4.23 shows a 3×3 square filled in, one number at a time. The goal is to fill the cells with consecutive whole numbers so that the sum of the numbers in every row, every column, and both diagonals is 15. There are several ways to accomplish this, but they are all reflections or rotations of the same basic idea. The first number, 1, is placed in the middle of the top row.

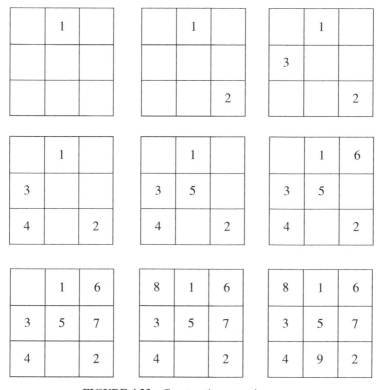

FIGURE 4.23 Constructing a magic square.

If we think of the 3×3 grid as a small map, with the usual north–south and east–west orientations, the whole grid can be filled in successfully by this strategy:

Rule 1: If possible, place the next number in the cell immediately to the northeast of the last number. This rule requires that there is an empty cell immediately to the northeast, but there are two potential pitfalls—the current cell might be at an edge of the grid so that moving northeast takes us outside the grid, and the cell immediately to the northeast might already have a number in it.

Rule 2: If we are at an edge, wrap around to the other side. Think of the grid as having its top connected to its bottom, and its right side connected to its left side.

Rule 3: If the cell immediately to the northeast already has a number, place the next number instead in the cell immediately to the south.

The number 1 was placed in a traditional location, the center of the top row.

2 was placed according to rule 2, wrapping around. If the bottom row was transferred so that it was above the top row, the path from 1 to 2 would be northeast.

3 was also placed according to rule 2, wrapping around. If the left column was transferred so that it was alongside the right column, the path from 2 to 3 would be northeast.

4 was placed according to rule 3, moving south because the cell to the northeast was already occupied.

5 was placed according to rule 1.

6 was placed according to rule 1.

7 was placed according to rule 3, since the wraparound protocol of rule 2 would have taken us to the cell already occupied by the number 4.

8 was placed according to rule 2.

9 was placed according to rule 2.

The natural reaction to seeing the implementation of this process is to object that rule 1 is hardly ever used, because moving to the northeast is constantly thwarted by edges and occupied cells. But rule 1 is the primary rule for filling magic squares by this technique, and on larger grids, like 5×5, 7×7, and 9×9, a greater and greater proportion of the cells will be filled according to rule 1.

You can verify that in the final grid in Figure 4.23 each row across, down or diagonally, adds up to 15. If you are familiar with the moves of the chess pieces bishop and knight, you will observe that the numbers 1-2-3 and 7-8-9 trace a knight's moves, while 4-5-6 is a bishop's move. The common difference from cell to cell in these moves is 1. You will also observe that the numbers 1-4-7 and 3-6-9 trace a knight's moves, while 2-5-8 is a bishop's move. The common difference from cell to cell in these moves is 3. We will see these chess piece patterns again.

A second way to construct a 3×3 magic square is to put three arbitrary numbers in three cells and use small algebraic equations to force other numbers into the other six cells. This technique is illustrated in the six steps of Figure 4.24, where the numbers 7, 10, and 12 were chosen arbitrarily. The 12 and the 10 are in corners, and the 7 is in a side cell opposite those corners. The location of the numbers is much more important than the actual value of the numbers. This technique would work with *any* choices of numbers but with only *several* similar choices of cell locations.

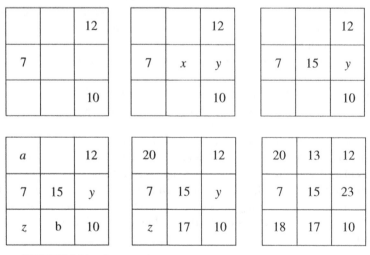

FIGURE 4.24   Constructing a magic square by a different technique.

In the second step of Figure 4.24 two cells have been filled in as **x** and **y**. Because the sums of all rows and columns have to be equal in magic squares, we can say that

$$7 + \mathbf{x} + \mathbf{y} = 12 + \mathbf{y} + 10$$

even though we don't yet know what those sums are. We can subtract **y** from both sides of this equation, to get

$$7 + \mathbf{x} = 12 + 10$$

This forces **x** to be 15, and in the third step of Figure 4.24 we see the 15 filled in.

Three additional cells are marked in the next step of Figure 4.24: **a**, **b**, and **z**. There is a diagonal $(12 + 15 + \mathbf{z})$ that must have the same sum as the bottom row $(\mathbf{z} + \mathbf{b} + 10)$, and this tells us, after discounting **z**, that $12 + 15 = \mathbf{b} + 10$, so **b** has to be 17.

Similarly, $\mathbf{a} + 7 + \mathbf{z} = 12 + 15 + \mathbf{z}$, so $\mathbf{a} + 7 = 12 + 15$, and $\mathbf{a} = 20$.

The fifth step of Figure 4.24 shows **a** and **b** replaced by their calculated values, and now the grid has so many numbers filled in that we know what the sum of every row and column must be. The northwest-to-southeast diagonal is $20 + 15 + 10$, or 45. To make the magic square "magic", **y** must be 23 and **z** must be 18, and the cell in the middle of the top row must be 13.

Let's look at this finished 3×3 magic square with respect to the gimmick of knight's moves and bishop's moves that we saw in the previous magic square. The patterns hold. Look at the sequences 7-12-17, 13-18-23, and 10-15-20, all with a "jump" of $+5$ from cell to cell, and at the sequences 7-10-13, 17-20-23, and 12-15-18, all with a "jump" of $+3$ from cell to cell.

There are many more patterns and regularities to be found in magic squares of various sizes. The first technique we saw with the 3×3 magic square (always striving to put the next number upward and to the right, wrapping around when necessary, and dropping down one square when wrapping around is impossible) works for any square that has an odd number of cells to fill.

PROBLEM 4.1   Complete this 5×5 magic square, using the same rules employed in the 3×3 magic square earlier.

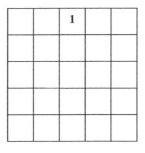

PROBLEM 4.2   Complete this 7 × 7 magic square, using the same rules.

			1			

The second technique we saw is limited to 3 × 3 magic squares.

PROBLEM 4.3   Complete this 3 × 3 magic square.

		5
3		
		14

PROBLEM 4.4   Complete this 3 × 3 magic square.

31		
		27
14		

There are a substantial number of ancient Chinese "story" problems, involving arithmetic, algebra, and geometry, where the story puts the calculations in a fanciful context. Here are some examples.

> A farmer is buying birds. A rooster costs 5 coins. A hen costs 3 coins. And for 1 coin he can buy 3 chickens. He has 100 coins and desires 100 birds.
> **How many of each does he buy?**

This of course is not exactly a real-life situation, and, additionally, someone not terribly familiar with poultry might wonder what exactly is the difference between a hen and a chicken. No matter. Our concern is finding a solution.

You might observe that some combinations give us the same number of birds as coins.

1 hen and 3 chickens (4 birds) cost 4 coins.
1 rooster and 6 chickens (7 birds) cost 7 coins.

This is the key to solving problems like this. Notice that we need to combine some number of birds that are more expensive and cost more than one coin per bird (hens or roosters) and some number of birds that are less expensive and cost less than one coin per bird (chickens).

If we can buy 4 birds for 4 coins, we can buy 8 birds for 8 coins, 12 for 12, and so on. And if we can buy 7 birds for 7 coins, we can 14 for 14, 21 for 21, and so on.

If we can carefully pick multiples of 4 and 7 that add up to 100, we'll have a solution. In fact, we'll find several solutions because there are several combinations of multiples of 4 and 7 that add up to 100.

$$100 + 0 = 100 \quad 0 \text{ roosters} \quad 25 \text{ hens} \quad 75 \text{ chickens}$$
$$72 + 28 = 100 \quad 4 \text{ roosters} \quad 18 \text{ hens} \quad 78 \text{ chickens}$$
$$44 + 56 = 100 \quad 8 \text{ roosters} \quad 11 \text{ hens} \quad 81 \text{ chickens}$$
$$16 + 84 = 100 \quad 12 \text{ roosters} \quad 4 \text{ hens} \quad 84 \text{ chickens}$$

The answer given by the Chinese author is 4 roosters, 18 hens, and 78 chickens, and this choice makes the most practical sense: a farmer needs a few roosters, but not too many.

It is worth noticing that the differences between solutions involve 28 and multiples of 28, and 28 is the product of 4 and 7, the numbers we found when we were looking for small numbers of birds and coins that match. The different solutions essentially substitute 4 groups of 7 birds for 7 groups of 4 birds. You can buy 28 birds for 28 coins two different ways: 4 roosters and 24 chickens, or 7 hens and 21 chickens.

PROBLEM 4.5   The teacher wants to give gifts to his 100 students. For his favorite students, he will buy more expensive gifts, but he insists on at least giving something to every student.

Exquisite gifts cost $6 apiece.
Nice gifts cost $2 apiece.
Cheap gifts are sold at three for a dollar.

The teacher will spend exactly $100 and buy gifts of each type. How many of each does he buy?

PROBLEM 4.6   A total of 100 hungry people, who happen to have exactly $100 among them, decide to go to a fast-food restaurant for lunch. The menu offers three distinct options:

1. **A "meal fit for a king" costs $7.**
2. **The "regular meal" costs $3.**
3. **The "peasant plate" is cheap: two portions are sold for $1.**

They want to spend exactly $100 and get exactly 100 meals. Somehow, they will determine among themselves which people get the best meals, but your task is to find a few options for them, as far as how many of which type of meal they can buy.

The dog is chasing the rabbit, and the distance between them is 50 yards. They run at constant speeds. The dog will catch the rabbit after the rabbit runs 125 yards.

Once the distance between them has narrowed to 30 yards, how much further will the dog have to run?

This is a problem in rates. While it isn't stated explicitly in the problem that the rabbit is running *away from* the dog, we can safely assume that it is. The dog will cover 175 yards ($125 + 50$) in total. Therefore, the dog runs 7 yards for every 5 yards the rabbit runs.

If r is how many yards the rabbit will run before it is caught once the gap is reduced to 30 yards, the dog will run $r + 30$ yards, and the proportion

$$\frac{7}{5} = \frac{r+30}{r}$$

will give us the answer, 105 yards. (This is the *dog's* distance; remember: $r = 75$, but r is the *rabbit's* distance.) The proportion is set up so that both fractions express the dog's distance in the numerator and the rabbit's distance in the denominator.

We could also solve by observing that for the dog's distance, which we can call d,

$$\frac{30}{50} = \frac{d}{175}$$

The gap between the dog and the rabbit starts at 50 yards and is eventually reduced to zero (when the dog catches the rabbit). Since both animals run at constant speeds, the gap between them shrinks at a constant rate. When the gap is down to 30 yards, the dog still has $\frac{30}{50}$ (or $\frac{3}{5}$) of his running to do. $\frac{3}{5}$ of 175 yards is 105 yards.

The cistern has two inlet taps used to fill it and one outlet tap used to empty it. The cistern can hold 48,000 gallons when filled to capacity.

If the cistern is empty and only tap #1 is opened, the cistern will fill in 12 hours.
If the cistern is empty and only tap #2 is opened, the cistern will fill in 6 hours.
If the cistern is full and only tap #3 is opened, the cistern will empty in 8 hours.

Assuming the cistern is empty and all three taps are opened, how long will it take to fill the cistern completely?

The key to solving this problem is figuring out what happens in 1 hour. The first tap will fill the 48,000 gallons in 12 hours, so it lets in 4000 gallons per hour. A similar division tells us the second tap lets in 8000 gallons per hour. And the third tap releases 6000 gallons per hour. The net hourly flow is 12,000 gallons in and 6,000 gallons out, or a combined net flow of 6,000 gallons in. Dividing the capacity (48,000) by the hourly net flow tells us that the cistern fills in 8 hours when all three taps are opened.

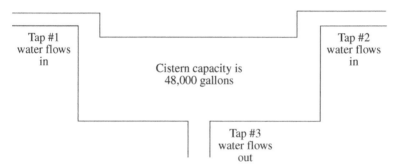

FIGURE 4.25 A Chinese problem involving rates. Water flows into the cistern, and simultaneously water flows out of the cistern.

You can confirm your understanding of the last problem by solving the next problems.

PROBLEM 4.7   A cistern has two inlet taps used to fill it and one outlet tap used to empty it. The cistern can hold 90,000 gallons when filled to capacity.

If the cistern is empty and only tap #1 is opened, the cistern will fill in 18 hours.
If the cistern is empty and only tap #2 is opened, the cistern will fill in 15 hours.
If the cistern is full and only tap #3 is opened, the cistern will empty in 10 hours.

Assuming the cistern is empty and all three taps are opened, how long will it take to fill the cistern completely?

PROBLEM 4.8   A cistern has two inlet taps used to fill it and one outlet tap used to empty it. The cistern can hold 84,000 gallons when filled to capacity.

If the cistern is empty and only tap #1 is opened, the cistern will fill in 4 hours.
If the cistern is empty and only tap #2 is opened, the cistern will fill in 6 hours.
If the cistern is full and only tap #3 is opened, the cistern will empty in 3 hours.

Assuming the cistern is empty and all three taps are opened, how long will it take to fill the cistern completely?

Some men are in the marketplace to buy a quantity of chickens. If each man chips in $9, they have $11 more than they need. If each man chips in $6, they have $16 less than they need.
   How many men are in the group, and what is the price of that quantity of chickens?

You might use a guess-and-check strategy to see what would happen with a certain number of men, but the more efficient technique is to make a system of equations. The little story can be expressed in terms of the two unknown quantities. Let's call D the total number of dollars that the men need, and let's call M the number of men. We can translate the scenarios as follows:

$$9M = D + 11$$
$$6M = D - 16$$

It is a simple matter now to eliminate either variable by subtraction or by substitution. There are several equally good ways to do this.

One way is to add 27 to both sides of the second equation.

$$6M \qquad\quad = \quad D - 16$$
$$6M \;+\; 27 \;=\; D + 11$$

Now we have two distinct expressions that equal $D + 11$:

$$9M \qquad\quad = \quad D + 11$$
$$6M \;+\; 27 \;=\; D + 11$$

So, $9M$ must equal $6M + 27$ (they are both equal to $D + 11$).

Subtracting $6M$ from both expressions gives us: $3M = 27$.

So M is 9. And then substituting back into either of the original equations tells us that D is 70.

It all makes sense. If the 9 men each contribute $9, they have $81, and that's $11 extra if the chickens cost $70. On the other hand, if the 9 men contribute only $6 each, they have $54, and they are $16 short of the $70 cost.

The idea in this problem, that in one scenario the men have too much money and in another scenario they have too little, echoes the "excess and deficiency" problems featured in one chapter of *Nine Chapters on the Mathematical Art*. That chapter presents the special Chinese variation of the *false position* algebra that we saw previously in Egyptian papyri problems. False position solving techniques were used also in India, the Arab world, and medieval Europe, but the unique Chinese twist was to make two guesses for the unknown quantity: one guess was intentionally too high (the excess) and one guess was intentionally too low (the deficiency), and the Chinese found a formula that combined the guesses and the outcomes of the guesses to determine the correct answer.

PROBLEM 4.9 **Some men are in the marketplace to buy a quantity of rice. If each man chips in $22, they have $10 more than they need. If each man chips in $17, they have $35 less than they need.**

How many men are in the group, and what is the price of that quantity of rice?

PROBLEM 4.10 **Some men have decided to buy a fine horse. If each man chips in $52, they have $26 less than they need. If each man chips in $55, they have $10 more than they need.**

How many men are in the group, and what is the price of the horse?

A town is in the shape of a circle, with a straight north–south street as its diameter, gates at each end of that street, and a circular wall 15 feet high. If you walk out of the town through the *north* gate and continue walking straight, you will find a cherry tree exactly 135 yards north of the town. A man walks out of the *south* gate for 15 yards and then turns and walks west until he can see the cherry tree with his view unobstructed by the town wall. This requires him to walk 208 yards west.

How long is the diameter of the town?

This is a complicated set-up, and solving it algebraically involves working with a quartic equation, one with a number raised to the fourth power. Apparently, the Chinese were not terribly inconvenienced by this. But to understand this problem fully, we need a diagram and a small lesson on similar triangles.

The set-up of the town and the cherry tree and the man's line of sight is:

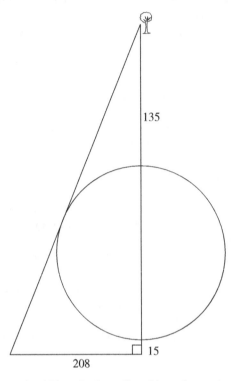

FIGURE 4.26   The town is within a circular wall, and its main street runs north-south through its center. The challenge in this Chinese word problem is to find the diameter of the town, given how far the man has to walk to be able to see the cherry tree.

A key consideration is that the man has to walk as far as 208 yards west, because if he were to walk any smaller distance his view of the cherry tree to the north would be obstructed by the high town walls. So we have a right triangle where one side is the diameter of the town (or twice the radius) plus 135 plus 15, and one side is 208. The Pythagorean theorem gives us the hypotenuse, in terms of r.

The hypotenuse is $\sqrt{(2r+150)^2 + 208^2}$

Now, to determine the diameter of the town, we need a second triangle, but before we find that triangle, let's do the little lesson on similar triangles.

We say two triangles are similar if they have exactly the same three angles. This forces them to have the same shape, and it makes the ratio of any pair of their corresponding sides a constant. Consider the two sets of triangles in figure 4.27, which differ only slightly.

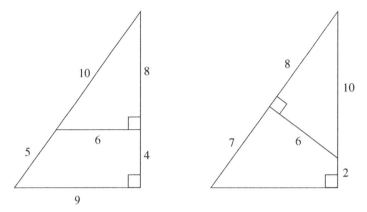

FIGURE 4.27 These triangles illustrate the key concept behind the solution to the Chinese problem with the circular town.

On the left side, we have a triangle with sides of length 9, 12, and 15. We can see this is a right triangle, as $9^2 + 12^2 = 15^2$. A line of length 6 is drawn parallel to the base of 9, creating a second triangle with the same three angles. This line breaks the 12 into lengths of 8 and 4, and it breaks the 15 into lengths of 10 and 5, and it also outlines a smaller triangle within the original triangle, and this smaller triangle has sides of 6, 8, and 10. You can verify that the 6-8-10 triangle is also a right triangle, since $6^2 + 8^2 = 10^2$.

The ratio of corresponding sides of these triangles is $3:2$. The particular values are $9:6$ (the short sides adjacent to the right angle), $12:8$ (the long sides adjacent to the right angle), and $15:10$ (the hypotenuses), and they all can be reduced to $3:2$.

On the right side of Figure 4.27, we also see a 9-12-15 triangle, but the line of length 6 is situated differently. It is drawn so that it meets the big hypotenuse at a $90°$ angle, splitting the 15 into lengths of 7 and 8, and splitting the side of length 12 into lengths of 10 and 2. You can use the $180°$ rule to verify that the triangles have the same angles. And you can verify that the small triangle on the right is identical to the small triangle on the left, even though it is oriented differently.

The ancient Chinese know about the Pythagorean theorem and they knew about similar triangles. They also knew that a tangent line meets a circle at a $90°$ angle. They used all this knowledge to essentially rig the "town and cherry tree" problem, to make the answer come out evenly as a whole number. Let's add a line to the diagram on the previous page.

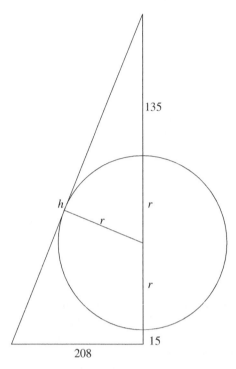

FIGURE 4.28 Another radius has been added to the overview of the town. This radius creates the similar triangles that lead to the solution.

This new line is a radius of the circle, and since it meets the big hypotenuse at a right angle, it creates a small right triangle that is similar to the big right triangle. We can describe the sides of the triangles in terms of the radius of the town and the big hypotenuse:

	Short Side	Middle Side	Hypotenuse
Big triangle	208	$2r+150$	h
Small triangle	r	Part of hypotenuse	$r+135$

We can make a proportion and say that

$$\frac{208}{r} = \frac{h}{r+135}$$

Crossmultiplying gives us $(r)\cdot(h)=(208)\cdot(r+135)$, and since h can be expressed in terms of r, we can make an equation with just one unknown, r.

This is not easy to solve, because it involves the expression r^4 (r to the 4th power), but the Chinese had their methods and they did solve it. Interestingly, equations with expressions like r^3 and r^4 really stumped European mathematicians, and it was not until late Medieval and early Renaissance times that some Italian geniuses worked out the general methods.

EXERCISE

Verify (by the Pythagorean theorem and by the ratios of the sides of the triangles) that the diameter of the town is 240 yards.

It seems very likely that the ancient Chinese composer of this problem began by figuring out the lengths of the sides of two large similar right triangles. Then he oriented the smaller triangle to have one side perpendicular to the larger triangle's hypotenuse, and then he crafted the story about the street and the gates and the cherry tree so that the dimensions worked out just right.

Let's see how the composer might have worked out the quartic equation. The hypotenuse of the large triangle is

$$\sqrt{(208)^2 + (2r+150)^2}$$

so the equation $(r) \cdot (h) = (208) \cdot (r+135)$ can be reexpressed as follows:

$$(r) \cdot \sqrt{(208)^2 + (2r+150)^2} = (208) \cdot (r+135)$$

To get rid of the radical (our notation, not theirs), we square both sides:

$$r^2 \cdot \left[(208)^2 + (2r+150)^2 \right] = (208)^2 \cdot (r+135)^2$$

Let's square those big numbers:

$$r^2 \left[43,264 + 4r^2 + 600r + 22,500 \right] = (43,264)\left(r^2 + 270r + 18,225 \right)$$

If the terms on the left side are combined and distributed, and the terms on the right side are distributed, the result is

$$4r^4 + 600r^3 + 65,764r^2 = 43,264r^2 + 11,681,280r + 788,486,400$$

You can see already that the Chinese must have been comfortable with enormous numbers. A small simplification is possible, subtracting $43,264r^2$ from both sides:

$$4r^4 + 600r^3 + 22,500r^2 = 11,681,280r + 788,486,400$$

Another simplification is to divide both sides by 4, since every term is a multiple of 4:

$$r^4 + 150r^3 + 5,625r^2 = 2,920,320r + 197,121,600$$

The Chinese were phenomenally good at factoring. The term r^2 can obviously be factored out of the left side, but less obviously $5625 = 75^2$. The large numbers on the right side are multiples of 24^2 (576) and 13^2 (169), so a huge simplification of the previous equation is

$$(r)^2 (r+75)^2 = 97,344(30r+2025)$$

We can take the square root of both sides and get

$$(r)(r+75) = 312\sqrt{(30r+2025)}$$

This doesn't look so nice, but the Chinese had extensive knowledge of squares and square roots and would have been able to find that this last equation is true when $r=120$.

How could they do that? Their mathematicians recognized squares and factors and divisors and kept good records of which numbers were divisible by which factors. Let's think of a smaller number and how you might recognize its factors. Consider the number 225. Since it ends in 5, you probably realize it's divisible by 5. Since its digits add up to a multiple of 3 ($2+2+5=9$), you probably realize it's divisible by 3. So, what happens when you divide 225 by 5 and by 3? You get 15, which itself is 5×3, so you understand that $225 = 15^2$. The Chinese would have observed that 97,344 is 312^2 and that $312 = 13 \times 24$, and they would have realized that for the dimensions of the circular town to work out nicely a few things had to be true: ($30r+2025$) had to be a perfect square, and the factors (r) and ($r+75$) had to be multiples of 13 and 24. This limited the numbers they had to investigate.

If r is 13, then ($r+75$) is 88, which is not a multiple of 24. If r is 24, then ($r+75$) is 99, which is not a multiple of 13. Eventually they would have found that r could be 120. 120 is a multiple of 24, and $120+75 = 195$, which is a multiple of 13. Of course, if $r=120$, the diameter of the town is 240 yards.

Let's try a few very similar problems where the numbers are not so monstrous.

PROBLEM 4.11   A camp is in the shape of a circle, with a straight north–south path as its diameter, gates at each end of that path, and a circular wall 12 feet high. If you walk out of the camp through the *north* gate and continue walking straight, you will find an oak tree exactly 9 yards north of the camp. A man walks out of the *south* gate for 20 yards and then turns and walks west until he can see the oak tree with his view unobstructed by the camp wall. This requires him to walk 24 yards west. How long is the diameter of the camp?

PROBLEM 4.12   A camp is in the shape of a circle, with a straight north–south path as its diameter, gates at each end of that path, and a circular wall 9 feet high. The commander walks out of the camp through the *north* gate and continue walking north for another 25 yards. A soldier walks out of the *south* gate and continues walking south for 56 yards before turning west and walking until he can see the commander with his view unobstructed by the camp wall. This requires him to walk 36 yards west. How long is the diameter of the camp?

At the market, oranges come in boxes of 3 different sizes. Chang wants to buy exactly 100 oranges, and he observes that the following combinations all contain exactly 100 oranges: 2 large boxes and 1 medium box, 3 medium boxes and 1 small box, or 4 small boxes and 1 large box.
   What is the capacity of each size?

This is basically a "system of equations" problem. There are 3 unknowns (the capacities of the large, medium, and small boxes) and 3 equations. Today, this problem would be solved quickly using matrix algebra, but matrix algebra wasn't invented when the Chinese came up with this brain teaser. We'll solve for one variable in terms of the others and substitute, to find the solution.

The letters L (large), M (medium), and S (small) can represent the capacities of the boxes, and we can interpret the problem to get these equations:

$$
\begin{aligned}
2L + M &= 100 \\
3M + S &= 100 \\
4S + L &= 100
\end{aligned}
$$

Our strategy is to replace M and S in the second equation with equivalent expressions for L, based on manipulating the first and third equations.

The first equation tells us that $M = 100 - 2L$.

The third equation tells us that $S = \dfrac{100 - L}{4}$

If we put these equivalent expressions into equation two, replacing M and S, we get

$$
3(100 - 2L) + \frac{100 - L}{4} = 100
$$

Multiply by 4 to get rid of the denominator:

$$
12(100 - 2L) + (100 - L) = 400
$$

Distribute:

$$
1200 - 24L + 100 - L = 400
$$

Collect like terms:

$$
1300 - 25L = 400
$$

Subtract 400 from both sides:

$$
900 - 25L = 0
$$

Add $25L$ to both sides:

$$
900 = 25L
$$

Divide by 25:

$$36 = L$$

Substituting this result for L into the first equation gives

$$2(36) + M = 100$$

So $M = 28$.
Substituting this result for L into the third equation gives

$$4S + 36 = 100$$

So $S = 16$.
You can check that the second equation is good too: $3(28) + 16 = 100$.

PROBLEM 4.13   A cook wants to measure out exactly 56 ounces of water, and he has two ladles of different sizes. Call the ladles A and B.

Three scoops of ladle A are not quite enough.
Also, 5 scoops of ladle B are not quite enough.

But the following combinations work precisely:

3 scoops of ladle A plus 1 scoop of ladle B
5 scoops of ladle B plus 1 scoop of ladle A

What are the capacities of the two ladles?

PROBLEM 4.14   A cook wants to measure out exactly one gallon of water, and he has three ladles of three different sizes. Call the ladles A, B, and C.

Two scoops of ladle A are not quite enough.
Also, 3 scoops of ladle B are not quite enough.
Even 4 scoops of ladle C are not quite enough.

But the following combinations work precisely:

2 scoops of ladle A plus 1 scoop of ladle B
3 scoops of ladle B plus 1 scoop of ladle C
4 scoops of ladle C plus 1 scoop of ladle A

What are the capacities of the three ladles?

Next we have a geometry problem, which is presented as simply a geometry problem, with no fanciful story attached. This also has to be solved using knowledge of similar triangles.

A right triangle has sides of 5 and 12 meeting at the 90° angle.
   What is the length of the side of the largest square that can fit inside the triangle?

There are two ways to draw a square inside a 5-12-13 right triangle. The square can nestle against the right angle, with one corner touching the hypotenuse, or the square can run alongside the hypotenuse, with corners touching the smaller sides. We will see that the first way produces a square with a side slightly longer than 3.5 units, and the second way produces a square with a side slightly shorter than 3.5 units.

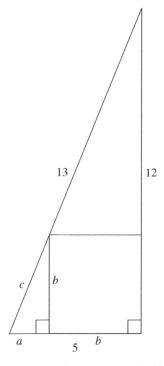

FIGURE 4.29   There are two ways to situate a square inside the right triangle. Here the square shares a 90° angle with the triangle.

In Figure 4.29, the square nestles against the right angle. Coincidentally, the 5-12-13 triangle is split into three parts: the square and two small triangles that are similar to the 5-12-13 triangle. We know all the triangles are similar because of the angles they share, as a consequence of two sides of the square being parallel to sides of the 5-12-13 triangle.

We can label the sides of the smallest triangle a, b, and c, where $a^2 + b^2 = c^2$, and make two very clear statements about a and b:

$$\text{Statement 1}: \quad (a + b) = 5$$

$$\text{Statement 2}: \quad \frac{a}{b} = \frac{5}{12}$$

The thing of most interest to us is the length of b, because b is the side of the square, and we can use a system of equations to find b simply. Rearranging the first statement gives

$$b = 5 - a$$

And multiplying this by 12 gives

$$12b = 60 - 12a \tag{4.1}$$

Crossmultiplying the second statement gives

$$5b = 12a \tag{4.2}$$

We have produced two new equations based on the original statements, and adding Equations 4.1 and 4.2 gives

$$17b = 60$$

So, $b = \dfrac{60}{17}$, or $3\dfrac{9}{17}$

This is the side of the largest square that can fit inside a 5-12-13 triangle, and it is slightly larger than 3.5.

EXERCISE

Show that the square formed in the 5-12-13 triangle by the second placement method (one side on the hypotenuse) has a side length of $\dfrac{720}{229}$ units, or 3 and $\dfrac{93}{229}$. This is slightly less than 3.5.

Use the relationships $\dfrac{a}{b} = \dfrac{5}{12}$ and $\dfrac{b}{c} = \dfrac{c}{5-a}$

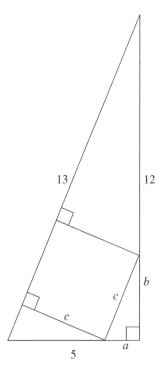

FIGURE 4.30 In this arrangement, one side of the square lies on the hypotenuse of the 5-12-13 triangle.

Finally, one of the great gems of ancient Chinese math is the procedure known as the Chinese Remainder Theorem, which is still used today in a branch of mathematics called "number theory." It is a systematic way to find large numbers by analyzing what happens when you divide them by small numbers. For example, we could pose this question:

A girl has a collection of dolls.

    If she sorts them, one at a time, into 5 piles, she has 4 dolls left over.

    If she sorts them, one at a time, into 3 piles, she has 1 doll left over.

How many dolls does she have?

Let's consider the scenario with 5 piles. Since she's putting equal numbers of dolls into each pile, the total number of dolls in piles will be a multiple of 5. These are numbers like 0, 5, 10, 15, 20, and so on. But she has 4 dolls left over when she does this, so her total number of dolls will be 4 more than a multiple of 5. She might have 4, 9, 14, 19, 24, and so on dolls.

Let's consider the scenario with 3 piles. Since she's putting equal numbers of dolls into each pile, the total number of dolls in piles will be a multiple of 3. These are numbers like 0, 3, 6, 9, 12, and so on. But she has 1 doll left over when she does this, so her total number of dolls will be 1 more than a multiple of 3. She might have 1, 4, 7, 10, 13, and so on dolls.

We are looking for numbers that appear on both of these lists:

4   9   14   19   24   29   34   39   44   49   54   59   64   69   74   79   84   ...

1   4   7   10   13   16   19   22   25   28   31   34   37   40   43   46   49   ...

Numbers on both lists are 4, 19, 34, and 49, and you might have observed that the gap between pairs of these numbers is 15. That might lead you to guess that 64 is the next common number that will come along, followed by 79 and 94, and you would be right.

4, 19, 34, 49, 64, 79, and 94 are all correct answers to the question *how many dolls could she have*? In number theory, these answers can be expressed by a formula. The girl has **4 + 15k** dolls, where $k = \{0, 1, 2, 3, \ldots\}$ and this means that when we put zero or any positive integer into the formula as the value of k, we get a number of dolls that satisfies the conditions.

We could take a different approach to get to the same conclusion. Because there is a remainder of 4 when the dolls are divided into 5 piles, we could say the number of dolls equals **5x + 4**, where $x = \{0, 1, 2, 3, \ldots\}$. You can confirm that this formula generates the row of numbers above that starts with 4 and 9.

Because there is a remainder of 1 when the dolls are divided into 3 piles, we could say the number of dolls equals **3y + 1**, where $y = \{0, 1, 2, 3, \ldots\}$. You can confirm that this formula generates the row of numbers above that starts with 1 and 4.

Since there are these two ways to express the same numbers of dolls, we are looking for situations where **5x + 4 = 3y + 1**. We can do a bit of algebra to find those situations.

Starting with

$$5x + 4 = 3y + 1$$

subtract 4 from both sides:

$$5x = 3y - 3$$

and factor:

$$5x = 3(y - 1)$$

and now divide both sides by 5:

$$x = \frac{3}{5}(y - 1)$$

This is a linear equation with two variables, x and y, and you may recall that this kind of equation has an infinite number of solutions: you pick a value for y, plug it into the formula, and it generates a value for x. Since, in this problem, the girl has a whole number of dolls, we are interested in values of y that lead to integer values of x. $(y-1)$ has to be a multiple of 5 in order to cancel the denominator and make the right side of the equation a whole number, so $y = \{6, 11, 16, 21, 26, \ldots\}$.

Let's try the first value out: $y = 6$, so $(y-1) = 5$, and $\dfrac{3}{5}$ of that is 3, and $x = 3$. We go back to the expression for the number of dolls, $5x + 4$, and plug in 3 and see that the number of dolls is 19. Good.

Let's try the second value out: $y = 11$, so $(y-1) = 10$, and $\dfrac{3}{5}$ of that is 6, and $x = 6$. We go back to the expression for the number of dolls, $5x + 4$, and plug in 6 and see that the number of dolls is 34. Also good.

So we have two methods for finding multiple solutions to problems of this type. We can find two sequences of numbers and look for matches between the sequences, or we can develop an algebraic expression that characterizes the variables. As long as the numbers involved are relatively small, the first method is not too tedious, but with very large numbers or with more than two sets of divisors and remainders, the algebraic approach is definitely the better option. Before we look at complications of that type, why not try a couple of problems.

PROBLEM 4.15 A rich man has a collection of gold coins. If he sorts them into 8 piles, he has 3 left over. If he sorts them into 5 piles, he has 2 left over.
How many gold coins does he have?

PROBLEM 4.16 A rich man has a collection of silver coins. If he sorts them into 6 piles, he has 2 left over. If he sorts them into 7 piles, he has 4 left over.
How many silver coins does he have?

In general, problems like this require the solver to find the minimum solution or a specific solution tied to the story. So you could be asked the smallest number of coins possible, or a number of coins in a range in between two numbers, for example.

The Chinese Remainder Theorem tells us first whether a problem like this has an answer. (It might not.) Then it tells us something about the size of the smallest answer. Then it tells us what number we have to add (repeatedly) to the smallest answer to find all the additional answers.

A situation where a similar problem has no answer arises when the remainders are illogical. If dividing into 10 piles leaves a remainder of 2, dividing into 5 piles cannot leave a remainder of 3.

EXERCISE

Why is it impossible for a number to have a remainder of 2 when it is divided by 10, and a remainder of 3 when it is divided by 5?

Why is it impossible for a number to have a remainder of 3 when it is divided by 9, and a remainder of 2 when it is divided by 6?

And, really, so far we haven't covered exactly what the Chinese Remainder Theorem is. Because it's awfully complicated. Students majoring in mathematics in college generally study number theory after they've studied calculus, and just to express the Chinese Remainder Theorem accurately requires statements about modular arithmetic, congruence, and numbers being relatively prime to each other, and it involves some notation that would baffle anyone who is not a math major. The Chinese Remainder Theorem (besides asserting that there is a solution to the kind of problems we've been looking at, and saying something about the nature of the solution) in its extended form describes the process for finding that solution. What we've been looking at so far is just the groundwork that led to the Chinese Remainder Theorem.

The Chinese Remainder Theorem is approximately 2000 years old, but the ideas that led to it (what we've looked at) are probably 1000 years older.

Let's see what happens when we make the problem more difficult by adding a third divisor.

A girl has a collection of dolls.

If she sorts them into 5 piles, she has 3 dolls left over.
If she sorts them into 4 piles, she has 1 doll left over.
If she sorts them into 3 piles, she has 2 dolls left over.

How many dolls does she have?

The numbers that make sense for the 5 pile split are as follows:

$$3 \quad 8 \quad 13 \quad 18 \quad 23 \quad 28 \quad 33 \quad 38 \quad 43 \quad 48 \quad 53 \quad 58, \text{ etc.}$$

The numbers that make sense for the 4 pile split are as follows:

$$1 \quad 5 \quad 9 \quad 13 \quad 17 \quad 21 \quad 25 \quad 29 \quad 33 \quad 37 \quad 41 \quad 45, \text{ etc.}$$

The numbers that make sense for the 3 pile split are as follows:

$$2 \quad 5 \quad 8 \quad 11 \quad 14 \quad 17 \quad 20 \quad 23 \quad 26 \quad 29 \quad 32 \quad 35, \text{ etc.}$$

It looks like it would be very tedious to hunt for numbers that are on all three lists. But we can express the number of dolls three ways: $5x+3$, $4y+1$, and $3z+2$. If we put the first two expressions together, we get

$$5x + 3 = 4y+1$$
$$5x = 4y-2$$

and finally

$$x = \frac{4y-2}{5}$$

To make x a whole number, $y = \{3, 8, 13, 18, 23, 28, \ldots\}$

If we put the last two expressions together, we get

$$3z + 2 = 4y + 1$$
$$3z = 4y - 1$$

and finally

$$z = \frac{4y - 1}{3}$$

To make z a whole number, $y = \{1, 4, 7, 10, 13, 16, 19, \ldots\}$

Now we have two lists to compare (instead of three), and it's easy to find the first match: 13. When $y = 13$, x and z are whole numbers (10 and 17), and the number of dolls is 53.

The Chinese Remainder Theorem tells us that additional solutions are in the following form:

$$\mathbf{53 + 60k} \quad \text{where } k = \{0,1,2,3,\ldots\}$$

and the reason 60 is so special is that $5 \times 4 \times 3 = 60$, and sorting dolls into 5, 4, and 3 piles means dividing a large number by 5, 4, and 3.

PROBLEM 4.17 A rich man has a collection of jewels. If he sorts them into 7 piles, he has 1 left over. If he sorts them into 6 piles, he has 5 left over. If he sorts them into 4 piles, he has 3 left over.

How many jewels does he have?

PROBLEM 4.18 A child has a collection of seashells. If he sorts them into 10 piles, he has 3 left over. If he sorts them into 9 piles, he has 7 left over. If he sorts them into 7 piles, he has 4 left over.

How many seashells does he have?

The Chinese Remainder Theorem is more and more helpful the larger the numbers we have to deal with. Let's imagine a scholar was challenged to find a number that, when divided by 44 left a remainder of 21, and, when divided by 103, left a remainder of 16. The kinds of lists we have seen so far would be extremely tedious to produce.

The scholar might represent the unknown number as $(44x + 21)$ and $(103y + 16)$ and generate a relationship like those we have seen already

$$x = \frac{103y - 5}{44}$$

Finding appropriate values for y (that cause x to be a whole number) looks difficult, but the scholar would have two big clues. First, y would have to be an odd number, to force the numerator of the fraction to be even, since an odd numerator divided by an even denominator cannot be a whole number. Second, the Chinese Remainder Theorem would suggest that there is an answer less than the product of 44×103, or 4532.

In fact, the scholar would find rather quickly (to his relief) that when $y = 15$, $x = 35$. Substituting back into $(44x + 21)$ and $(103y + 16)$ tells him that the original unknown number was 1561.

And here is where the Chinese Remainder Theorem works so wonderfully: if the scholar needed to find larger solutions, he would know immediately that the next answer is 6093, because 6093 is $1561 + 4532$. And the next one is 10,625 $(6093 + 4532)$.

Suggestions for book or Internet research	
History, archeology	Ancient China
	The Silk Road
	Shang dynasty
Mathematical topics	Ancient Chinese mathematics
	Traditional Chinese numbers
	The Chinese Remainder Theorem
	Counting boards
	Magic squares
	The Nine Chapters on the Mathematical Art
	Double false position algebra
Mathematicians	Sun Tzu
	Liu Hui

ANSWERS TO PROBLEMS

4.1

17	24	1	8	15
23	5	7	14	16
4	6	13	20	22
10	12	19	21	3
11	18	25	2	9

4.3

18	25	5
3	16	29
27	7	14

4.5

The key is to find combinations of gifts where the number of gifts equals the number of dollars. Let's call such a combination a "package."

A package of 5 gifts that costs $5 consists of 2 nice gifts and 3 cheap gifts.

A package of 11 gifts that costs $11 consists of 1 exquisite gift, 1 nice gift, and 9 cheap gifts.

A package of 17 gifts that costs $17 consists of 2 exquisite gifts and 15 cheap gifts.

Now the goal is to find combinations of 5, 11, and 17 that add up to 100.

One solution is 5 packages of 17 and 3 packages of 5 $(5 \times 17) + (3 \times 5) = 100$.

In this solution, there are 10 exquisite gifts, 6 nice gifts, and 84 cheap gifts.

Another solution is 5 packages of 11 and 9 packages of 5 $(5 \times 11) + (9 \times 5) = 100$.

In this solution, there are 5 exquisite gifts, 23 nice gifts, and 72 cheap gifts.

4.7

The first tap will fill the 90,000 gallons in 18 hours, so it lets in 5,000 gallons per hour. A similar division tells us the second tap lets in 6000 gallons per hour. And the third tap releases 9000 gallons per hour. The net hourly flow is 11,000 gallons in and 9,000 gallons out, for a combined net flow of 2,000 gallons in. One more division tells us that the cistern fills in 45 hours when all three taps are opened.

4.9

The equation

$$22M = D + 10$$

represents $22 per man equaling the full price of the rice, plus $10.

Similarly,

$$17M = D - 35$$

represents $17 per man equaling $35 less than the full price of the rice.

There are several ways to proceed. One simple way is to subtract the second equation from the first equation. This gives

$$5M = 45$$

which tells us that $M=9$. Substitution into either of the first two equations gives us $D=188$. So 9 men went together to buy $188 worth of rice.

4.11

We can draw the camp as a circle with radius r. The north–south path is $2r$. The distance from the oak tree at the north and the spot to the south where the man turns west is $9 + 2r + 20$, or $2r + 29$. A right triangle can be drawn with sides of $(2r + 29)$ and 24. A line connecting the center of the camp to this hypotenuse creates a smaller similar right triangle, and we can describe the sides of the triangles in terms of the radius of the camp and the big hypotenuse:

	Short Side	Middle Side	Hypotenuse
Big triangle	24	$2r + 29$	h
Small triangle	r	Part of hypotenuse	$r + 9$

We can make a proportion and say that

$$\frac{24}{r} = \frac{h}{r+9}$$

Crossmultiplying gives $(r)\cdot(h) = (24)\cdot(r+9)$, and since h can be expressed in terms of r, we can make an equation with just one unknown, r.

The hypotenuse of the large triangle is $\sqrt{(24)^2 + (2r+29)^2}$, so the equation $(r)\cdot(h)=(24)\cdot(r+9)$ can be reexpressed as follows:

$$(r)\sqrt{(24)^2 + (2r+29)^2} = (24)(r+9)$$

To get rid of the radical, we square both sides:

$$(r)^2 \left[(24)^2 + (2r+29)^2 \right] = (24)^2 (r+9)^2$$

Let's compute those numbers:

$$(r)^2 \left[576 + 4r^2 + 116r + 841 \right] = (576)(r^2 + 18r + 81)$$

If the terms on the left side are combined and distributed, and the terms on the right side are distributed, the result is

$$4r^4 + 116r^3 + 1417r^2 = 576r^2 + 10,368r + 46,656$$

Subtracting $576r^2$ from both sides:

$$4r^4 + 116r^3 + 841r^2 = 10,368r + 46,656$$

Factoring on both sides leads to

$$(r)^2 (2r+29)^2 = 5184(2r+9)$$

We can take the square root of both sides and get

$$(r)(2r+29) = 72\sqrt{(2r+9)}$$

To make this come out nicely, $(2r+9)$ has to be a perfect square, and since $72 = 8 \times 9$, between (r) and $(2r+29)$ one of them has to be a multiple of 8 and the other has to be a multiple of 9, or the prime factors of 8×9 ($2 \cdot 2 \cdot 2 \cdot 3 \cdot 3$) have to be distributed in some combination between (r) and $(2r+29)$. $(2r+9)$ is a perfect square when $r=0$, $r=8$, $r=20$, and other values, and when we test them out we see how well $r=8$ works. The factors on the left become 8 (a multiple of 8), and 45 (a multiple of 9), and $(8)(45) = (72)\sqrt{25}$, so the path through the camp from the north gate to the south gate is 16, and the similar right triangles we created have sides of 8-15-17 and 24-45-51.

4.13

We can translate the problem into two equations:

$$3A + B = 56$$
$$5B + A = 56$$

Triple and rearrange the second equation:

$$3A + 15B = 168$$

Subtract the first equation from this new equation:

$$
\begin{array}{rcrcr}
3A & + & 15B & = & 168 \\
3A & + & B & = & 56 \\
\hline
& & 14B & = & 121
\end{array}
$$

This tells us $B=8$.
    Substituting into either of the original equations tells us $A=16$.

4.15

The number of coins can be described as $8x+3$ or as $5y+2$.
    Starting with

$$8x + 3 = 5y + 2$$

subtract 3 from both sides:

$$8x = 5y - 1$$

and divide both sides by 8:

$$x = \frac{5y-1}{8}$$

Since the rich man has a whole number of coins, we are interested in values of y that lead to integer values of x. $(5y-1)$ has to be a multiple of 8 in order to cancel the denominator and make the right side of the equation a whole number, so $(5y-1)$ has to be a number like 8, 16, 24, 32, 40, and so on, which means $5y$ has to be a number like 9, 17, 25, 33, 41, and so on, and any of these numbers that is a multiple of 5 will do.

    25 is the first appropriate number on the list, so if $y=5$, $5y=25$, $(5y-1)=24$, and $x=3$.

    The number of coins was described as $8x+3$ or as $5y+2$, and making the substitutions in both cases leads us to conclude that the smallest number coins the man could have is 27. Could he have more? Sure. And the Chinese Remainder Theorem tells us that the product of the divisors $(8 \times 5 = 40)$ is what we add to find additional solutions.

    Larger answers therefore are: 67, 107, 147, 187, and so on. You can verify that these quantities of gold coins also give the appropriate remainders when divided by 8 and by 5.

4.17

The numbers that make sense for the 7 pile split are as follows:

$$8 \quad 15 \quad 22 \quad 29 \quad 36 \quad 43 \quad 50 \quad 57 \quad 64 \quad 71 \quad 78 \quad 85, \text{ etc.}$$

The numbers that make sense for the 6 pile split are as follows:

$$11 \quad 17 \quad 23 \quad 29 \quad 35 \quad 41 \quad 47 \quad 53 \quad 59 \quad 65 \quad 71 \quad 77, \text{ etc.}$$

The numbers that make sense for the 4 pile split are as follows:

$$7 \quad 11 \quad 15 \quad 19 \quad 23 \quad 27 \quad 31 \quad 35 \quad 39 \quad 43 \quad 47 \quad 51, \text{ etc.}$$

It looks like it would be very tedious to hunt for numbers that are on all three lists.
But we can express the number of dolls three ways: $7x+1$, $6y+5$, and $4z+3$.
If we put the first two expressions together, we get

$$7x+1 = 6y+5$$
$$7x = 6y+4$$

and finally

$$x = \frac{6y+4}{7}$$

To make x a whole number, $y = \{4, 11, 18, 25, 32, \ldots\}$
If we put the last two expressions together, we get

$$4z+3 = 6y+5$$
$$4z = 6y+2$$

(dividing)

$$2z = 3y+1$$

and finally

$$z = \frac{3y+1}{2}$$

To make z a whole number, $y = \{1, 3, 5, 7, 9, 11, \ldots\}$
Now we have two lists to compare (instead of three) and it's easy to find the first match: 11. When $y = 11$, x and z are whole numbers (10 and 17), and the number of jewels is 71.

The Chinese Remainder Theorem tells us that additional solutions are in the following form:

$$\mathbf{71+84}k \quad \text{where } k = \{0,1,2,3,\ldots\}$$

and the reason 84 is so special is that $7 \times 6 \times 4 = 168$, and 84 is half of 168. The interval between consecutive solutions is the product of the problem's divisors if the divisors have no factors in common, but less than that product if the divisors share common factors. In this problem, two of the divisors, 6 and 4, are both multiples of 2, so we divide 168 by 2.

5

BABYLONIAN MATHEMATICS

As far as we can tell, Babylonian math is as old as Egyptian math and, actually, a good deal more sophisticated. And it's more complicated, in a way that baffles both students and historians. We, the Egyptians, and countless others, use a base-10 system where symbols and digits represent progressively larger powers of 10: 1, 10, 100, 1,000, 10,000, and so on. The Babylonians used a system where symbols and digits represent progressively larger powers of 60: (1), (60), (3,600), (216,000), (12,960,000), and so on.

So, a number like 325, which we break down as $(3 \times 100) + (2 \times 10) + (5 \times 1)$, they would read quite differently. The 3, which to us represents the number of hundreds, to them would represent the number of 3600s. The 2 would represent the number of 60s, and the 5 would represent (just as it does for us) the number of 1s. The full translation of the number is shown in Figure 5.1.

3	2	5
3 × 3600	2 × 60	5 × 1
10,800	120	5
Total = 10,925		

FIGURE 5.1 The Babylonian symbols "325" represent a number vastly bigger than 325 because of their base-60 system.

An Introduction to the Early Development of Mathematics, First Edition. Michael K. J. Goodman.
© 2016 John Wiley & Sons, Inc. Published 2016 by John Wiley & Sons, Inc.

Their notation was all based on triangular wedges, some pointing sideways and some pointing down, and small digits like 3, 2, and 5 would be written as follows:

▼▼▼   ▼▼   ▼▼▼▼▼

The Babylonians would have stacked the last bunch somewhat, to make

▼▼▼   ▼▼   ▼▼▼
                          ▼▼

It is very cumbersome for modern readers to constantly interpret the triangles, so in this book we will frequently use a common convention for representing Babylonian numbers. The brackets in $[3, 2, 5]_{60}$ alert us that we are considering a Babylonian base-60 quantity, and the commas between the digits tell us we have a 3-digit number. What makes base-60 difficult for so many people used to base-10 is that in base-60 there are 60 different digits. In base-10, we have 10 digits.

What do those other digits look like? To the Babylonians, they were combinations of horizontal wedges and vertical wedges, but we're mostly going to use the number symbols we are more familiar with. We just saw that $[3, 2, 5]_{60}$ is 10,925. Let's interpret $[31, 21, 51]_{60}$, a number with larger "digits."

In the same way that the 3 in $[3, 2, 5]_{60}$ told us there were three 3600s, the 31 in $[31, 21, 51]_{60}$ tells us there are thirty-one 3600s. The entire translation is shown in Figure 5.2.

31	21	51
31 × 3600	21 × 60	51 × 1
111,600	1206	51
Total = 112,911		

FIGURE 5.2   How the Babylonian number $[31, 21, 51]_{60}$ is understood to mean 112,911 in our numbers.

The Babylonians would have written 31 as ►►►▼ where the three sideways pointing triangles represent tens. (For convenience, I use the symbol ► for ten. The actual cuneiform is more like an open triangle pointing in the opposite direction.)

Their 21 would have been ►►▼, and their 51 would have been ►►►►►▼.

So the Babylonian version of 112,911 would have looked something like this:

►►►▼   ►►▼   ►►►►►▼

There in a nutshell is the problem students have with Babylonian math: you have to make sense of the triangular wedges, and you have to get used to base-60.

We are so used to doing arithmetic in base-10 that we hardly notice the assumptions we make and the conventions we follow. To do base-60 arithmetic, we have to follow the same conventions, but allow for all the extra digits and the vastly larger place values from digit to digit. When we add numbers today, sometimes there is "carrying" and sometimes there isn't, depending on whether our subtotal in any column exceeds 9. When we subtract numbers, sometimes there is "borrowing" and sometimes there isn't, depending on whether the digit we are subtracting exceeds the digit above it. Similar rules apply in base-60.

$$\begin{array}{r} [5,13]_{60} \\ + \quad [3,28]_{60} \\ \hline [8,41]_{60} \end{array} \qquad \begin{array}{r} [5,33]_{60} \\ + \quad [3,28]_{60} \\ \hline [9,1]_{60} \end{array}$$

In the first example, there is no carrying, because $13+28=41$, and 41 is a valid digit in base-60. In the second example, there is carrying, because $33+28=61$, and 61 is not a valid digit in base-60. The highest valid digit in base-60 is 59, and we have to represent 61 as $[1, 1]_{60}$, one 60 plus one 1. (You may want to verify, by translating, that the first example says $313+208=521$. And that the second example says $333+208=541$.) Let's look at the corresponding subtractions.

$$\begin{array}{r} [8,41]_{60} \\ - \quad [3,28]_{60} \\ \hline [5,13]_{60} \end{array} \qquad \begin{array}{r} [9,1]_{60} \\ - \quad [3,28]_{60} \\ \hline [5,33]_{60} \end{array}$$

In the first subtraction, there is no borrowing. $41-28=13$ and $8-3=5$, making the columns very easy for us. In the second subtraction, we're momentarily flummoxed by the problem of taking 28 away from 1. So we borrow. In our regular base-10 arithmetic, when we borrow, we reduce the digit on the left by one and add 10 to the digit on the right. We do this because the digit on the left represents quantities 10 times larger than the digit on the right. In base-60 arithmetic, when we borrow, we reduce the digit on the left by one and add 60 to the digit on the right. We do this because the digit on the left represents quantities 60 times larger than the digit on the right. Borrowing in this example momentarily changes the problem to

$$\begin{array}{r} [8,61]_{60} \\ - \quad [3,28]_{60} \end{array}$$

and we can see why the answer $[5, 33]_{60}$ makes sense.

You might already see that base-60 numbers, while they make look like base-10 numbers, are much, much larger than the base-10 numbers they look like. $[4, 5, 6, 7]_{60}$ (see Figure 5.3) looks like 4567, but it represents $882,367$.

$[5, 1, 8, 6, 2]_{60}$ (see Figure 5.4) looks like 51,862 but represents $65,045,162$.

$[13, 7, 22, 56, 29]_{60}$ (see Figure 5.5) looks awfully strange because most of its base-60 digits look like 2-digit numbers to us. It represents $170,074,589$.

The Babylonians didn't use the brackets and the commas—those are modern notations meant to simplify our translations. The Babylonians only used the wedges. And this introduced some ambiguity. Depending on where the downward-pointing

4	5	6	7
4 × 216,000	5 × 3,600	6 × 60	7 × 1
864,000	18,000	360	7
Total = 882,367			

FIGURE 5.3   Translating $[4, 5, 6, 7]_{60}$.

5	1	8	6	2
5 × 12,960,000	1 × 216,000	8 × 3,600	6 × 60	2 × 1
64,800,000	216,000	28,800	360	2
Total = 65,045,162				

FIGURE 5.4   Translating $[5, 1, 8, 6, 2]_{60}$.

13	7	22	56	29
13 × 12,960,000	7 × 216,000	22 × 3,600	56 × 60	29 × 1
168,480,000	1,512,000	79,200	3360	29
Total = 170,074,589				

FIGURE 5.5   Translating $[13, 7, 22, 56, 29]_{60}$.

wedge was, it might represent 1 or 60, or 3600, or more, or even less, like one-sixtieth. The earliest Babylonians didn't have a decimal indicator or a zero. (Later Babylonians eventually got around to using a pair of diagonally oriented wedges to represent zero, and this eliminated much of the ambiguity.)

So the wedges and arithmetic baffle students. What baffles historians? They wonder: why in the world did the Babylonians use base-60 in the first place?

Nobody knows definitively. There are explanations that involve how easy it is to divide 60 evenly: it is divisible by 2, 3, 4, 5, 6, 10, 12, 15, 20, and 30. But other numbers are also easily divided. There are explanations that involve measurements of time, like 60 seconds in a minute and 60 minutes in an hour. But those divisions of time actually came later in history. There are explanations that involve geometry and the 360° in a circle, 360 being an obvious multiple of 60. But again, the convention of measuring angles in degrees came later. There are explanations that tie Babylonian units of measure to earlier Sumerian units of measure, but of course we could simply push the mystery further into the past and ask where those Sumerian ideas originated.

(It would be as if we marveled at our own comfort with 12 inches equaling 1 foot, and 3 feet equaling 1 yard, and therefore based our number system on powers of 36: 1, 36, 1,296, 46,656.) There are calendar explanations: the cycle of the moon is close to 30 days (half a 60), and the cycle of the earth around the sun is close to 360 days (six 60s). But these are rather imprecise estimates for a civilization that was so sophisticated in its other measurements and calculations. There are explanations that propose that base-60 was a compromise or combination of earlier systems, like base-5 and base-12. Some primitive people count in base-5, using the fingers on one hand only. Others have a counting system based on the knuckles of the four fingers opposite the thumb, and there are 12 finger joints when you count this way. But again, why would a sophisticated society adopt a clumsy compromise?

One pretty compelling idea is that the core of the Babylonian system is base-10, and that the Babylonians counted by tens until they got to 60 because 60 was a "convenient" number in their minds. Think about our currency and coins. Since we are a base-10 society, it might be natural to assume that our money would include only one dollar bills, ten dollar bills, hundred dollar bills, thousand dollar bills, and so on. In fact we have five dollar bills and twenty dollar bills, because 5 and 20 are "convenient" numbers for our numerical purposes. We even have a rarely used two dollar bill, and a fifty dollar bill. Among our coins, we naturally have the penny and the dime (base-10 landmarks), but we also have the nickel and the quarter and the lesser-used fifty cent piece. The nickel and the quarter are "convenient" amounts. Perhaps, in some way we don't see from our perspective, thousands of years later, 60 was regarded as a "convenient" number.

And there is also the possibility that 60 achieved its prominence for an arbitrary reason that was then solidified by tradition. Why is a baseball game nine innings long? Why is a tennis score announced as "love fifteen" at the first point? Why does a golf course have 18 holes? Why do we split the pound into 16 ounces, and split the gallon into 4 quarts? How did exactly 5280 feet become a mile? When major league baseball expanded in the middle of the 20th century, the number of games in the regular season increased from 154 to 162—where did those numbers come from?

Our calendar has two very sensible numbers that approximate our observations of the sun and the moon. The number of days in the year and the number of days in a month are tied to the sun and the moon. But what about the week? Is 7 days some unit that occurs in nature? Absolutely not. It is entirely a man-made tradition. At various times in history, idealists and utopians have tried to change it, probably most famously during the French revolution when the 10-day week was briefly in vogue. If we discovered that the Babylonians used a 6-day week, that might lend some support to the special status of 60, but in fact we know that the Babylonians and Sumerians used a 7-day week and named their days for the sun, the moon, and the five planets they could see. Our 7-day week is ultimately derived from theirs.

Maybe it grew out of their notation. The number 59 is represented

where the wedges pointing sideways are full "tens" and the wedges pointing down are nine of the next "ten." Maybe some important scribe just said *enough is enough* and decided to make 60 look a lot simpler.

The historians of math have had a good time with this puzzle, and they have some intriguing material to work with. A related number system, from the ancient city of Mari in what is now eastern Syria, used different cuneiform patterns of wedges, and the Mari system was base-10. Clay tablets have been translated that reveal the units of measure the Babylonians used for weight, volume, size, distance, and value. Records of earlier, later, and contemporary people in Mesopotamia have been analyzed, so that Sumerian, Akkadian, Assyrian, Hittite, and Chaldean records can be compared to Babylonian records. Still, there is no universally accepted explanation for base-60.

With respect to base-60, however, the archeological record is incredibly rich. Around the world, in university museums and national museums, collections include hundreds of thousands of cuneiform tablets that were excavated at dozens of sites in Mesopotamia. Cuneiform is the wedge-shaped writing system the Babylonians used. Using a sharp stick, they notched their alphabetical characters and their numerals into soft clay, and the clay was heated until it hardened, and the tablets basically last forever.

The tablets show us the calculations Babylonian students made in what we might term "schoolbook" problems, and they show us complex, sophisticated, theoretical mathematical ideas as well. But before we look at this math, we ought to get more comfortable with the cuneiform wedges and carrying out the simple mathematical operations in base-60.

The Babylonians basically had one symbol for a digit, ▼, whereas we have ten (0, 1, 2, 3, 4, 5, 6, 7, 8, 9), so the orientation and spacing were critical. Otherwise ambiguous interpretations would make the system far too confusing. Look at these 3 numbers, which all have the elements for 30 (three horizontal triangles) and 6 (six vertical ones):

►►►▼▼▼ ▼▼▼	►►► ▼▼▼ ▼▼▼	►►► ▼▼▼ ▼▼▼		
Looks like one digit	Looks like two separate digits	Looks like two separate digits		
	Narrow space in between the clusters of wedges	Wide space in between the clusters of wedges		
		A zero is implied in between the two clusters, making a third digit		
30 + 6	30 × 60 6	30 × 3600	0 × 60	6
36	1806	108,006		

FIGURE 5.6 Empty space between groups of wedges differentiated between similar-looking Babylonian numbers. It was a weakness of their system that they did not have a character for zero, since the person reading the number might guess incorrectly what the empty space meant.

We can write these numbers in a bracket notation, understanding that they are base-60.

$$36 \text{ is simply} [36] \quad 1,806 \text{ is } [30,6] \quad 108,006 \text{ is } [30,0,6]$$

It is highly recommended that you patiently examine other examples of Babylonian numbers and practice interpreting them.

PROBLEM 5.1   Express this number, written in the Babylonian system, as a modern number:

PROBLEM 5.2   Express this number, written in the Babylonian system, as a modern number:

Working in the other direction, taking one of our numbers and expressing it in base-60 or with the cuneiform wedges, requires breaking the number into pieces that are multiples of the powers of 60. To convert 785, for example, we start by noticing that 785 is more than 60 but less than 3600. This makes 60 the largest power of 60 that divides into 785 a whole number of times. So we compute that 785 divided by 60 is 13 with a remainder of 5, and thus we write $[13, 5]_{60}$ to represent 785.

Let's examine the conversion of a larger number: 99,697. We start by observing that 99,697 is more than 3,600 but less than 216,000, and this tells us that 3,600 is the largest power of 60 that divides into 99,697 a whole number of times. We compute that $\dfrac{99,697}{3600} = 27$ with a remainder of 2497. Our base-60 number will therefore start with 27 as its first digit. The remainder 2497 is between 60 and 3600, so we continue the process by dividing 2497 by 60. This gives 41 with a remainder of 37. And this tells us that the base-60 equivalent of 99,697 is $[27, 41, 37]_{60}$. You can check that this is correct by translating $[27, 41, 37]_{60}$ according to the method already described.

PROBLEM 5.3   Show how the ancient Babylonians would have represented the number 8533.

PROBLEM 5.4   Show how the ancient Babylonians would have represented the number 830,844.

Babylonian scribes would have had no particular reason to translate numbers into their base-10 equivalents and vice versa, but they would have had to be experts in base-60 arithmetic. Here is an addition problem that might have been a typical exercise: add

$$[21,35,17]_{60} + [15, 44, 52]_{60}$$

Of course, this problem would have been expressed as something like

We start with $17 + 52$, the last digits. That's 69. But 69 is more than 60, so we "carry" the 60 by adding 1 to the middle digits and write 9 as the last digit of our result.

We now have $35 + 44 + 1$ in the middle. $35 + 44 + 1$ is 80, and 80 is more than 60.

So we carry the 60 by adding 1 to the first digits, leaving 20 as the middle digit of our answer.

The first digit is now the sum $21 + 15 + 1$, or 37. So our final answer is $[37, 20, 9]_{60}$. You can use your calculator to check this result.

$$[21, 35, 17]_{60} \text{ is } (21 \times 3600) + (35 \times 60) + 17, \text{ or } 77,717.$$
$$[15, 44, 52]_{60} \text{ is } (15 \times 3600) + (44 \times 60) + 52, \text{ or } 56,692.$$
$$77,717 + 56,692 = 134,409.$$
$$[37, 20, 9]_{60} \text{ is } (37 \times 3600) + (20 \times 60) + 9, \text{ or } 134,409. \text{ It matches.}$$

Here's a subtraction problem: $[2, 5]_{60} - [1, 23]_{60}$

You can confirm that this is $125 - 83$.
So we expect the answer to be 42.

Breaking it down step by step, our first step should be to subtract 23 from 5.
We'll have to "borrow" from the next column over.
So, $[2, 5]$ becomes $[1, 65]$, momentarily an invalid number.
But we can subtract 23 from 65, giving 42 as the last digit, and then 1 minus 1 makes the first digit zero: $[0, 42]$. We wouldn't even write the leading zero, so our answer would simply be $[42]_{60}$.

The "carrying" and "borrowing" in base-60 are just like what you learned in elementary school in base-10, except that the numbers are a lot bigger. You need to collect 60 before you can "carry." When you "borrow," you get 60.

One way to do Babylonian style subtraction is to represent the numbers as wedges, cross out all equivalent wedges, and show the answer as the remaining wedges. The previous problem, $125 - 83$, is shown in Figure 5.7.

PROBLEM 5.5   Add $[55, 30, 19]_{60}$ and $[25, 34, 49]_{60}$.

PROBLEM 5.6   Add $[40, 8, 12, 39]_{60}$ and $[3, 56, 11, 40]_{60}$.

125		83
▼▼ ▼▼▼▼▼	The numbers are shown as wedges. On the left, we see 2 × 60, and 5 ones. On the right, we see one 60, and then 23.	▼ ►►▼▼▼
[▼]▼ ▼▼[▼▼▼]	Brackets show equivalent wedges that can be cancelled from both sides.	[▼] ►►[▼▼]
▼ ▼▼	Here is what's left after the cancelling. We've taken away one 60 and 3 ones.	►►
60 2		20
►►► ▼▼ ►►►	The 60 is now converted into 6 tens.	►►
►►► ▼▼ ►[►►]	Brakets again show equivalent wedges that can be cancelled.	[►►]
►►► ▼▼ ►	42 is left on the left side. This is the answer.	

FIGURE 5.7   The subtraction problem $125-83$.

PROBLEM 5.7   Subtract $[1, 23, 20, 0]_{60}$ from $[2, 18, 53, 20]_{60}$.

PROBLEM 5.8   Subtract $[7, 44, 31, 25]_{60}$ from $[9, 18, 18, 6]_{60}$.

The student should practice converting our numbers to base-60, and converting base-60 numbers to our numbers, and adding and subtracting in the Babylonian style.

Multiplying and dividing in base-60 looks extremely complicated, particularly with the cuneiform numbers, and the Babylonians must have felt the same way, because they devised some shortcuts. There were two principal multiplication methods. They both relied on records of previous multiplications preserved on clay tablets. One method turned multiplication into addition, and the other method turned multiplication into addition, subtraction, and cutting in half. The shortcut for division was to turn it into multiplication.

Multiplication tables have been found in many archeological sites, making it appear that every scribe had a small library of them. A typical multiplication table looked like Figure 5.8 (in translated form).

This is a reference table for multiplying by 25. On the top half, we see the values for N from 1 to 5, and from 6 to 10. The row that has **4** and **1, 40** is expressing that 4×25 is 100, because **1, 40** represents one 60 and 40 ones. The row that has **9** and **3, 45** is expressing that 9×25 is 225, because **3, 45** represents three 60s and 45 ones.

N	$25 \times N$	N	$25 \times N$
1	25	6	2, 30
2	50	7	2, 55
3	1, 15	8	3, 20
4	1, 40	9	3, 45
5	2, 5	10	4, 10

N	$25 \times N$
20	8, 20
30	12, 30
40	16, 40
50	20, 50
60	25

FIGURE 5.8   A Babylonian table for multiplying by 25.

The lower half of the table shows values for N between 20 and 60. For example, 40×25 is 1000. In base-60, **16, 40** represents sixteen 60s and 40 ones.

The last entry on the table represents twenty-five 60s (1500 to us), and this shows one of the weaknesses of the Babylonian system, since it is exactly the same symbol we see on the first row for $N = 1$. In other words, the numbers 25 and 1500 are both represented by the cuneiform ▶▶▼▼▼▼▼.

But the table is very easy to use. Suppose a scribe needed to compute 34×25. He would find the values that corresponded to 30 and 4 and add them: $[12, 30] + [1, 40] = [14, 10]$

Suppose he needed to compute 73×25. He would find the values that corresponded to 60 and 10 and 3 and add them: $[25, 0] + [4, 10] + [1, 15] = [30, 25]$

PROBLEM 5.9   Construct a base-60 table for multiplying by 17. Show values for N from 1 to 10 and 20, 30, 40, 50, and 60.

PROBLEM 5.10   Construct a similar table for multiplying by 48.

A second type of multiplication table preserved on clay tablets was basically a list of integers and their squares. This was useful because the Babylonians were experts at algebra and had calculated a general formula for multiplication that relied on the squares of the numbers being multiplied.

You may recall a formula from your high school days that says $(a + b)^2 = a^2 + 2ab + b^2$. The middle term in this formula, **2ab**, is twice the product of two integers, **a** and **b**. So if you wanted to multiply **a** times **b**, you could express **a** times **b** in terms of this formula.

Reverse the two sides of the formula: $a^2 + 2ab + b^2 = (a + b)^2$

Subtract a^2 and b^2 from both sides: $2ab = (a + b)^2 - a^2 - b^2$

Divide both sides by 2: $\mathbf{ab} = \dfrac{(\mathbf{a}+\mathbf{b})^2 - \mathbf{a}^2 - \mathbf{b}^2}{2}$

Observe that the terms in the numerator of the fraction on the right are all squares.
Let's say a Babylonian scribe needed to compute 54×38. He could put 54 and 38 into the formula:

$$(54)(38) = \frac{(54+38)^2 - 54^2 - 38^2}{2}$$

$54 + 38 = 92$. He could find 92^2 on his tables. $\mathbf{92^2}$ is 8464. $\mathbf{54^2}$ is 2916. $\mathbf{38^2}$ is 1444. He could take these numbers and subtract: $8464 - 2916 - 1444 = \mathbf{4104}$. This last number gets divided by 2: $\dfrac{4104}{2} = \mathbf{2052}$.

It appears the Babylonians developed this technique from geometry, not algebra. Imagine a rectangle that was 54×38 units. On the left side of Figure 5.9, the rectangle stands alone. On the right side, a 54×54 square has been placed on top of the rectangle, and a 38×38 square has been placed at the side of the rectangle. A fourth shape, the original 54×38 rectangle turned on its side, has been added so that it and the other shapes now form a larger square with dimensions 92×92.

54 × 54	38 × 54
54 × 38	38 × 38

54 × 38

FIGURE 5.9   One Babylonian multiplication technique relied on seeing a rectangle as part of a larger square.

The Babylonians figured out that the area of the 54×38 rectangle was half of what was left after the smaller squares were subtracted from the big square.

PROBLEM 5.11   Use the Babylonian squaring method to calculate 23×8.

PROBLEM 5.12   Use the Babylonian squaring method to calculate 67×41.

There were also some multiplication shortcuts available to a clever scribe who got comfortable working with numbers. Multiplying small numbers by 2 meant doubling the number of wedges.

▶▶▼▼▼   ▶▼▼   times 2 was   ▶▶▼▼▼   ▶▼▼
                                                 ▶▶▼▼▼   ▶▼▼

If the doubling produced an invalid base-60 digit, the invalid digit had to be converted to a valid digit:

►►►▼▼▼	►►▼	times 2 became	►►►▼▼▼	►▼▼
			►►►▼▼▼	►▼▼

but this had to be	▼	▼▼▼	►▼▼	because the
written as		▼▼▼	►▼▼	doubling

produced too many ►'s.

Multiplying by 4 meant doubling the number of wedges twice.

Multiplying small numbers by 10 meant changing the wedges from ▼ to ►.

Multiplying by 60 meant simply moving the symbols over a little. It's like us adding a zero when we multiply by 10.

Multiplying by 30 meant multiplying by 60 and then cutting in half (dividing by 2).

Multiplying by 20 meant multiplying by 60 and then dividing by 3.

Multiplying by 15 meant multiplying by 60 and then dividing by 4.

Division was treated as multiplication by a reciprocal. When we divide by two, we multiply by one half. When we divide by ten, we multiply by one tenth. The Babylonians made tables of reciprocals, so that division problems could be turned into multiplication problems, and reciprocals are fractions, and since 60 is evenly divisible by so many numbers, it was easy to generate a lot of fractions with the cuneiform wedges.

The only big problem for us, looking at their work thousands of years later, is that they didn't use a decimal point (or its equivalent), so we have to guess or infer where the whole number ends and where the fraction begins. For our convenience, we'll insert a semicolon into the base-60 notation we have been using, to indicate what mathematicians call a "sexigesimal" point, and we'll use a zero at the start of fractions.

(Sexigesimal means $\dfrac{1}{60}$ in the same way that decimal means $\dfrac{1}{10}$.)

In base-10, we say $\dfrac{1}{2}$ is 0.5 because the place value after the decimal point is tenths, and 5 tenths is one half. The place value after the base-60 sexigesimal point is sixtieths, and 30 sixtieths is one half, so in base-60 we write one half as $[0; 30]_{60}$. To a Babylonian scribe, a fraction was a number of parts out of 60.

$\dfrac{1}{2}$ was 30 parts, and therefore $[0; 30]_{60}$.

$\dfrac{1}{3}$ was 20 parts, and therefore $[0; 20]_{60}$.

$\dfrac{1}{4}$ was 15 parts, and therefore $[0; 15]_{60}$.

$\dfrac{1}{5}$ was 12 parts, and therefore $[0; 12]_{60}$.

$\dfrac{1}{6}$ was 10 parts, and therefore $[0; 10]_{60}$.

$\dfrac{1}{10}$ was 6 parts, and therefore $[0; 6]_{60}$.

$\dfrac{1}{12}$ was 5 parts, and therefore $[0; 5]_{60}$.

$\dfrac{1}{15}$ was 4 parts, and therefore $[0; 4]_{60}$.

There was no inhibition about using exclusively unit fractions, so

$\dfrac{4}{5}$ was 48 parts, and therefore $[0; 48]_{60}$.

$\dfrac{3}{4}$ was 45 parts, and therefore $[0; 45]_{60}$.

$\dfrac{2}{3}$ was 40 parts, and therefore $[0; 40]_{60}$.

$\dfrac{3}{5}$ was 36 parts, and therefore $[0; 36]_{60}$.

$\dfrac{7}{12}$ was 35 parts, and therefore $[0; 35]_{60}$.

$\dfrac{2}{5}$ was 24 parts, and therefore $[0; 24]_{60}$.

$\dfrac{3}{10}$ was 18 parts, and therefore $[0; 18]_{60}$.

$\dfrac{4}{15}$ was 16 parts, and therefore $[0; 16]_{60}$.

Some divisors don't go into 60 evenly, but they do go into 3600. In the same way that we go to two decimal places to represent hundredths, they go to two sexigesimal places to represent 3600ths.

$\dfrac{1}{8}$ is $[0; 7, 30]_{60}$. This is $\dfrac{7}{60} + \dfrac{30}{3600}$, which you can verify is $\dfrac{450}{3600}$, or $\dfrac{1}{8}$. You might see this more easily if you start with the idea that $\dfrac{1}{8}$ is half of $\dfrac{1}{4}$. The base-60 representation of $\dfrac{1}{4}$ is $[0; 15]_{60}$ and if you want to cut that in half, you see that 15 cut in half is 7.5. The symbol $[0; 7, 30]_{60}$ is showing 7 and "half of the next unit," or $\dfrac{30}{60}$ of it. $\dfrac{1}{16}$ is $[0; 3, 45]_{60}$. It's half of $[0; 7, 30]_{60}$.

Let's see how to represent $\frac{5}{9}$ in base-60. The denominator, 9, does not go into 60 evenly, but 9 is a divisor of 3600. $\frac{1}{9}$ of 3600 is 400. So $\frac{5}{9}$ of 3600 is $\frac{2000}{3600}$. What we need is the base-60 representation of 2000. 2000 divided by 60 is 33, with a remainder of 20, so the base-60 fraction $\frac{5}{9}$ is $[0; 33, 20]_{60}$. You can verify with your calculator that $\frac{33}{60}$ plus $\frac{20}{3600}$ is $\frac{5}{9}$.

PROBLEM 5.13   **Show these fractions in base-60: $\frac{7}{10}, \frac{3}{20}, \frac{5}{8}$.**

PROBLEM 5.14   **Show these fractions in base-60: $\frac{9}{10}, \frac{9}{25}, \frac{1}{32}$.**

Let's imagine that a Babylonian student or scribe needed to divide 175 by 25. He would have interpreted the problem as $175 \times \frac{1}{25}$ and grabbed his handy multiplication table for $\frac{1}{25}$ (see Figure 5.10).

He would have recognized that 175, $[2, 55]_{60}$ to him, was equal to $(60 + 60 + 50 + 5)$.

He would then, using the table, have added $[2; 24] + [2; 24] + [2] + [0; 12]$, which would have given him his answer, 7.

N	$1/25 \times N$	N	$1/25 \times N$
1	0; 2, 24	6	0; 14, 24
2	0; 4, 48	7	0; 16, 48
3	0; 7, 12	8	0; 19, 12
4	0; 9, 36	9	0; 21, 36
5	0; 12	10	0; 24
N		$1/25 \times N$	
20		0; 48	
30		1; 12	
40		1; 36	
50		2	
60		2; 24	

FIGURE 5.10   A Babylonian table for multiplying by (1/25).

EXERCISE

How would the Babylonian student do the division problem $\dfrac{175}{25}$ if he didn't have access to any tables? Assume that his teacher expressed the problem as [2, 55] divided by [25]. Of course, this would have been presented as follows:

▼▼   ▶▶▶▼▼▼   divided by   ▶▶▼▼▼
　　 ▶▶ ▼▼   ▼▼

N	$12 \times N$	N	$12 \times N$
1	12	6	1, 12
2	24	7	1, 24
3	36	8	1, 36
4	48	9	1, 48
5	1, 0	10	2, 0
N		$12 \times N$	
20		4, 0	
30		6, 0	
40		8, 0	
50		10, 0	
60		12, 0	

N	$1/5 \times N$	N	$1/5 \times N$
1	0; 12	6	1; 12
2	0; 24	7	1; 24
3	0; 36	8	1; 36
4	0; 48	9	1; 48
5	1	10	2
N		$1/5 \times N$	
20		4	
30		6	
40		8	
50		10	
60		12	

FIGURE 5.11 A comparison of the multiplication tables for 12 and for (1/5). Since 12 and 5 are considered "reciprocals" in base-60 multiplication, the tables are almost identical.

The Babylonian concept of reciprocals made it unnecessary to maintain many multiplication tables for particular fractions, because the multiplication tables for corresponding whole numbers could be used. We think of reciprocals as numbers which, when multiplied, equal 1. We say 12 and $\frac{1}{12}$ are reciprocals. The Babylonians recognized numbers that multiply to 60 as reciprocals, so 12 and 5 would be reciprocals to them. Let's look at the multiplication tables for 12 and for the fraction $\frac{1}{5}$ in Figure 5.11.

Now, inspection of these tables shows that the really important parts are exactly the same in all corresponding locations. What is not the same is the punctuation we introduce, the zeroes, the commas, and the semicolons. Take those away, and the tables are exactly the same. This echoes the idea that in base-60, Babylonian style, without zeroes and place values, every number is indistinguishable from a number 60 times as much, or 60 times as little. So the Babylonian table for multiplying by 12 would also serve quite well for dividing by 5 (which is multiplying by $\frac{1}{5}$).

The multiplication table for $\frac{1}{5}$ also suggests how a resourceful scribe might solve the problem 175 divided by 25 without a table for $\frac{1}{25}$. As long as he had a table for $\frac{1}{5}$, or 12, he could apply it twice. The first time, he would see that $[2, 55]_{60} \cdot [0; 12]_{60}$ was the sum of the numbers opposite 60 (twice) and 50 and 5: $12 + 12 + 10 + 1$, or 35. The second time, he would use 35 and take the sum of the numbers opposite 30 and 5: $6 + 1$.

What we've seen so far is that the Babylonians had expertise with numerical operations, and it is natural to wonder what they did with all this expertise.

Let's look at the math on some of those clay tablets that were excavated at the sites of the great old Mesopotamian cities (Babylon, Uruk, Nippur, Nineveh, and many more) and preserved today in museums around the world. There are literally tens of thousands of mathematical tablets in these museums. Naturally, there are tablets in Iraq's national museum. The Ottoman Empire ruled Iraq for a time, so there are tablets in Turkey's national museum. Archeologists from England, France, and Germany dug up many of the tablets, so there are tablets in the British Museum, the Louvre, and German museums. American academic expeditions found thousands of tablets, so there are vast collections at the University of Chicago, at Yale, at Columbia, and at the University of Pennsylvania.

Figure 5.12 shows an intriguing diagram from a clay tablet: a trapezoid is split into two smaller trapezoids by a line parallel to its bases. The question posed is: if the two smaller trapezoids have exactly equal areas, how long is the parallel line?

A reasonable first reaction is that we're missing some key information here. The area of a trapezoid depends on its height, and we were not given any information about the height of any trapezoid—the original big one, or the one on the left with a long base of 17, or the one on the right with a short base of 7.

Still, we can see that there must be an answer. Imagine drawing a parallel line within the original big trapezoid at any random spot, and then shifting the parallel line to the right or to the left to make the area of one side increase while the area

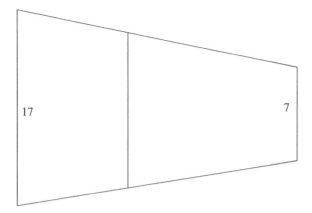

FIGURE 5.12   The big trapezoid is divided into two smaller trapezoids with equal areas by a line parallel to its bases. The Babylonian challenge was: how long is that line?

of the other side decreases. Somewhere there must be a line that divides the two sides evenly.

The Babylonian scribe who put this problem on a tablet knew that the height of the original trapezoid didn't matter, and that there was a brilliant shortcut for calculating the length of the parallel line. He knew to sum the squares of the two given bases, and divide that sum by 2, and then take the square root.

$$17^2 + 7^2 = 289 + 49 = 338$$
$$338 / 2 = 169$$
$$13^2 = 169$$

So the parallel line has a length of 13.

Observe that we still don't know the height of these trapezoids. The answer is 13 regardless of the heights!

In Figure 5.13, the large trapezoid from Figure 5.12 has been cut into smaller pieces, and we can use simple formulas for the areas of rectangles and triangles to verify that the areas of the two smaller trapezoids are equal. You may prefer to turn the diagram sideways to view the trapezoids as we traditionally do: with the parallel bases horizontal, but the algebra works regardless of the orientation.

We have the original bases of 17 and 7. The parallel line is 13. Horizontal lines have been added to split the 13 and the 17 into smaller segments. The distance between the 17 and the 13 is 10. The distance between the 13 and the 7 is 15.

The base with a length of 17 has been chopped into 5 segments, with lengths of 2, 3, 7, 3, and 2. The parallel line with a length of 13 has been chopped into 3 segments, with lengths of 3, 7, and 3.

So, the trapezoid on the left side of the parallel line consists of: a 7-by-10 rectangle, two 3-by-10 rectangles, and two 2-by-10 triangles; and the trapezoid on the right side of the parallel line consists of: a 7-by-15 rectangle and two 3-by-15 triangles. You can verify that the areas of the 5 shapes on the left are 70, 30, 30, 10, and 10 (totaling 150), and the areas of the 3 shapes on the right are 105, 22.5, and 22.5 (also totaling 150).

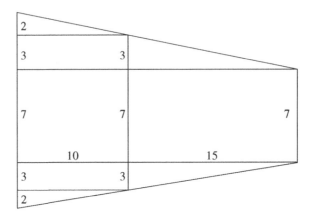

FIGURE 5.13 The big trapezoid is broken into rectangles and triangles.

Now, you may wonder where the heights of the trapezoids (10 and 15) came from. The formula for the area of a trapezoid is the average of the bases multiplied by the height, or

$$\frac{b_1 + b_2}{2} \cdot h$$

On the left side, the average of the two bases is 15, and on the right side the average is 10. To make the areas equal, we let each height equal the other trapezoid's average base.

It wasn't necessary to pick exactly 10 and 15 for the heights. As long as the ratio of the heights is the reciprocal of the ratio of the bases, the areas will be the same. That's why the height of the trapezoid wasn't specified in the original problem and why it wasn't necessary.

EXERCISE

In this longer trapezoid, with a height of 35, verify that the parallel line is also 13 and that the heights of the two smaller trapezoids are 14 and 21.

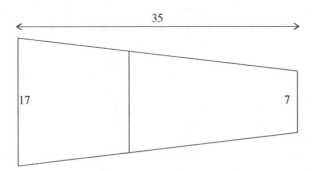

FIGURE 5.14 Another trapezoid with bases of 7 and 17 is split into two equal smaller trapezoids by a line of length 13.

Where did the Babylonian formula for the length of the parallel line (the square root of the sum the squares of the two bases divided by 2) come from? It came from a very nice piece of algebra based on two geometric ideas: the formula for the area of a trapezoid and the ratio of the sides of similar triangles. In Figure 5.15, S is the shorter base of the original trapezoid, L is the longer base, and P is the parallel line that splits the trapezoid into two smaller equal trapezoids. Small h is the height of the smaller trapezoid, which has S as one of its bases, and capital H is the height of the smaller trapezoid that has L as one of its bases.

The trapezoid on the left has an area of $\dfrac{P+S}{2} \cdot h$, and the trapezoid on the right has an area of $\dfrac{L+P}{2} \cdot H$, and these areas are equal.

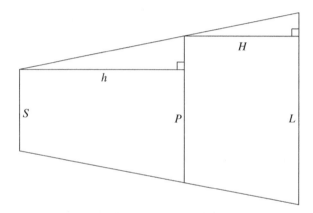

FIGURE 5.15   The generalized trapezoid.

The little triangles with bases H and h are similar, and we can compute the heights of those triangles in terms of S, P, and L. The height of the smaller triangle is $\dfrac{L-P}{2}$, and the height of the bigger triangle is $\dfrac{P-S}{2}$. These facts are enough to derive the remarkable Babylonian formula.

In the similar triangles, we can state this ratio among the sides:

$$\frac{h}{\dfrac{P-S}{2}} = \frac{H}{\dfrac{L-P}{2}}$$

Taking the reciprocal of both sides, and getting rid of the 2s, gives us the following:

$$\frac{(P-S)}{h} = \frac{(L-P)}{H}$$

We're going to save this result and use it in a moment. Meanwhile, we'll state that the areas of the two smaller trapezoids are equal:

$$\frac{P+S}{2} \cdot h = \frac{L+P}{2} \cdot H$$

Multiplying both sides of this equation by 2 yields this simplification:

$$(P+S) \cdot h = (L+P) \cdot H$$

Now it's time to use the result we saved. We can multiply its corresponding parts by the equation immediately above:

$$(P+S) \cdot h \cdot \frac{(P-S)}{h} = (L+P) \cdot H \cdot \frac{(L-P)}{H}$$

Now, rather conveniently, the h and H terms cancel out, leaving

$$(P+S)(P-S) = (L+P)(L-P)$$

which becomes

$$P^2 - S^2 = L^2 - P^2$$

Adding $(P^2 + S^2)$ to both sides to remove negative quantities gives

$$2P^2 = L^2 + S^2$$

and now we see where the Babylonian formula comes from. Solving this last equation for P, we find that P is the square root of half the sum of the squares of the bases. Remember: P is the parallel line that splits the original trapezoid evenly, and L and S are the bases of the original trapezoid.

On another tablet, the challenge is to *find two numbers that differ by 7 and multiply to 60*. The tablet shows the step-by-step solution to this problem, and following the process reveals to us the entire Babylonian approach to quadratic equations.

In modern terms, we would call the numbers x and $(x+7)$, and we would solve:

$$x(x+7) = 60$$
$$x^2 + 7x = 60$$
$$x^2 + 7x - 60 = 0$$
$$(x+12)(x-5) = 0$$

And $x = -12$ or 5. If $x = -12$, then $x+7$ is -5, and the product of -12 and -5 is indeed 60, which solves the problem. We have no problem with this solution, but the Babylonians did not use negative numbers. The second solution, $x = 5$, leads to the second number, $x+7$, being 12, and we and the Babylonians accepted this solution.

Our modern algebraic steps were to express the variables, form a product, distribute the x, subtract to set the expression to zero, factor, and solve.

$$\begin{array}{ll}
\text{Express the variables}: & x \text{ and } (x+7) \\
\text{Form a product} \quad : & x(x+7) = 60 \\
\text{Distribute} \quad : & x^2 + 7x = 60 \\
\text{Subtract} \quad : & x^2 + 7x - 60 = 0 \\
\text{Factor} \quad : & (x+12)(x-5) = 0 \\
\text{Solve} \quad : & x = -12 \text{ or } 5
\end{array}$$

The Babylonian method started by cutting the difference, 7, in half, to get 3.5
This was squared to get 12.25
This was added to the product, 60, to get 72.25
The square root of this was computed, 8.5
Half the difference, 3.5, was added and subtracted to this, giving 12 and 5, which we recognize as the positive solution for x.

The Babylonian procedure looks like some wild manipulation that just happened to luckily give us the right answer, but in fact the Babylonian method has a lot in common with our modern technique. While we called the numbers x and $(x+7)$, they in effect let x represent the number halfway in between the two numbers they were looking for. They solved

$$(x+3.5)(x-3.5) = 60$$

The two factors will of course differ by 7 (and multiply to 60). In our modern continuation of this modified version of the original problem, we will see all the numbers that turned up in the Babylonian method.

$$\begin{aligned}
(x+3.5)(x-3.5) &= 60 \\
x^2 + 3.5x - 3.5x - 12.25 &= 60 \\
x^2 - 12.25 &= 60 \\
x^2 &= 72.25 \\
x &= 8.5
\end{aligned}$$

Plugging 8.5 in for x in the equation $(x+3.5)(x-3.5) = 60$ makes the two factors 12 and 5.

The Babylonian formula relied on putting the quadratic equation in a format that we would recognize as $x^2 + Px = Q$ and computing the solution by using the formula in Figure 5.16.

In the problem *find two numbers that differ by 7 and multiply to 60*, P is the difference between the numbers (7), and Q is the result of the multiplication (60). The Babylonian procedure first finds half of P (3.5), then squares it (12.25), then adds Q (yielding 72.25). The square root of this (8.5) is reduced by half of P (3.5) to

$$x = \sqrt{\left(\frac{P}{2}\right)^2 + Q} - \frac{P}{2}$$

FIGURE 5.16   The Babylonian formula for solving quadratic equations expressed in the form $x^2 + Px = Q$.

give x (5). Since we are asked to find two numbers that differ by 7, the final answer is 5 and 12.

The formula for quadratics works perfectly for all numbers, but they only used it for positive numbers, as they apparently had no awareness of negative numbers. The Babylonian method forces us to express the quadratic in a very specific way: the square of a number, plus a multiple of that number, equals a constant. It works perfectly well once we've conformed to its rigid requirements.

Let's examine another quadratic equation solved in Babylonian style. They would solve $x^2 + 3x = 154$ by the following steps:

The factor 3, from the $3x$ term, is divided by 2, giving 1.5

This is squared, giving 2.25

This is added to the whole number 154, to give 156.25

The square root of this is computed, 12.5

The earlier result 1.5 is subtracted from this, giving 11.

11 is the answer.

Again, our modern approach would be to express the equation as $x^2 + 3x - 154 = 0$ and factor it. We would be looking for two factors of 154 that differed by 3, and eventually we would find 11 and 14. This would lead us to the equation

$$(x - 11)(x + 14) = 0$$

and we would accept both positive 11 and negative 14 as solutions.

PROBLEM 5.15   **Solve the following quadratic equation by both the modern and Babylonian methods:**

$$x^2 + 10x = 96$$

PROBLEM 5.16   **Solve the following quadratic equation by both the modern and Babylonian methods:**

$$x^2 + 18x = 1600$$

PROBLEM 5.17   **Solve by the Babylonian method:** $x^2 + 2x = 399$

PROBLEM 5.18   **Solve by the Babylonian method:** $x^2 + 15x = 1000$

EXERCISE

Show by algebraic manipulation that $x = \sqrt{\left(\dfrac{P}{2}\right)^2 + Q} - \dfrac{P}{2}$ **is always a solution for** $x^2 + Px = Q.$

Here is a wonderful result from a tablet:

```
        ▶▼▼▼        ▶▶
         ▼▼▼        ▶▶

        ▶▼▼▼        ▶▶
         ▼▼▼        ▶▶

  ▼▼    ▶▼▼▼      ▶▶▼▼▼      ▶▶
  ▼▼    ▶▼▼▼      ▶▶▼▼▼      ▶▶
         ▶▼
```

What could this mean?

$$[16, 40]$$

If we translate into base-60, the numbers are $[16, 40]$

$$[4, 37, 46, 40]$$

[16, 40] is 16 sixties plus 40, which is $960 + 40$, or 1000.

[4, 37, 46, 40] is $(4 \times 216,000) + (37 \times 3600) + (46 \times 60) + (40 \times 1)$, which looks pretty random, but it equals 1,000,000.

This tablet says $1000 \times 1000 = 1,000,000$

This is strong evidence that base-10 was fundamental to their mathematical thinking even though their notation was base-60. There could hardly be any reason to record a multiplication result like this unless powers of 10 were important to them.

The Babylonians had very good formulas for computing square roots and cube roots. In the problems we just looked at, it was critical to square numbers or find square roots or to do both, and if the numbers came out evenly, the Babylonians could have relied on their tables of roots and powers. But the numbers did not always come out evenly. The square root of 100 is 10, and the square root of 121 is 11, but every whole number between 100 and 121 does not have an obvious square root. But at least we can quickly say that the square root for any number between 100 and 121 is itself between 10 and 11. This is the foundation of the Babylonian formula for finding square roots.

To find the square root of 113, they expressed 113 as $10^2 + 13$. There were three key numbers for the Babylonians:

10, the largest whole number whose square was less than 113

$\dfrac{13}{2 \cdot 10}$, and

$\dfrac{13^2}{8 \cdot 10^3}$, or

$$10,$$

$$\frac{13}{20}, \text{ and}$$

$$\frac{169}{8000}$$

The calculation was $10 + \frac{13}{20} - \frac{169}{8000}$, which was 10.628875.

We know, from a quick tap on a calculator, that the actual square root is extremely close to 10.63015, so the Babylonian formula is not perfect. What, you may wonder, is 10.628875 squared? It is 112.973.

The Babylonians found this value close enough for their purposes, and since they had tables of squares and cubes of integers the calculation wasn't all that difficult for them.

The general Babylonian formula for square roots is shown in Figure 5.17, where A^2 and H add up to n.

Curiously, the Babylonian formula sometimes works better for larger numbers than for smaller numbers.

$$\sqrt{n} = \sqrt{A^2 + H} = A + \frac{H}{2A} - \frac{H^2}{8A^3}$$

FIGURE 5.17   The Babylonian formula for finding square roots.

The square root of 33, using a modern calculator, is about 5.74456

The Babylonian formula expresses 33 as $5^2 + 8$ and finds these three key numbers:

5, the largest whole number whose square was less than 33

$$\frac{8}{2 \cdot 5}, \text{ and}$$

$$\frac{8^2}{8 \cdot 5^3}, \text{ or}$$

$$5,$$

$$\frac{8}{10}, \text{ and}$$

$$\frac{64}{1000}$$

and finds 5.736 as the square root. 5.736 is 0.00856 less than the true answer.

The square root of 333, using a modern calculator, is about 18.24829

The Babylonian formula expresses 333 as $18^2 + 9$ and finds these three key numbers:

18, the largest whole number whose square was less than 333

$\dfrac{9}{2 \cdot 18}$, and

$\dfrac{9^2}{8 \cdot 18^3}$, or

18,

$\dfrac{9}{36}$, and

$\dfrac{81}{46656}$

and finds 18.24826 as the square root. 18.24826 is only 0.00003 less than the true answer.

The explanation for the greater accuracy for larger numbers is connected to the size of the numbers in the expression $A^2 + H$. The square root of A^2 is of course A, so the terms in the Babylonian formula that are added to and subtracted from A (the terms that involve H) can be regarded as adjustments when the starting number, N, is not a perfect square. The smaller the H is in comparison to A^2, the better the Babylonian formula is, since the adjustments are relatively smaller.

PROBLEM 5.19   Use the Babylonian formula to find the square root of 153. Use your calculator to see how close the Babylonian answer is to the actual value.

PROBLEM 5.20   Use the Babylonian formula to find the square root of 365.

PROBLEM 5.21   Use the Babylonian formula to find the square root of 2000.

PROBLEM 5.22   Use the Babylonian formula to find the square root of 7000.

The Babylonian formula for the cube root of a number also depends on expressing the number as the sum of two numbers. One of them is the largest cube less than that number. In this formula, n is the number, and A^3 and H add up to n.

$$\sqrt[3]{n} = \sqrt[3]{A^3 + H} = A + \frac{H}{3A^2} - \frac{H^2}{9A^5}$$

FIGURE 5.18   The Babylonian formula for finding cube roots.

The cube root of 113, the Babylonians saw, had to be between 4 and 5, because 4^3 is 64 and 5^3 is 125. They would express 113 as $4^3 + 49$ and find these three key numbers:

4, the largest whole number whose cube was less than 113

$$\frac{49}{3 \cdot 4^2}, \text{ and}$$

$$\frac{49^2}{9 \cdot 4^5}, \text{ or}$$

4,

$$\frac{49}{48}, \text{ and}$$

$$\frac{2401}{9216}$$

and find 4.7603 as the cube root. The calculator tells us 4.8346 is the cube root of 113, so the Babylonian formula, while okay, is not as precise as we'd expect.

But this is related to how big 49 is, relative to 64. Let's compute the cube root of 153. We can express 153 as $125 + 28$, or $5^3 + 28$, and find these three key numbers:

5, the largest whole number whose cube was less than 153

$$\frac{28}{3 \cdot 5^2}, \text{ and}$$

$$\frac{28^2}{9 \cdot 5^5}, \text{ or}$$

5,

$$\frac{28}{75}, \text{ and}$$

$$\frac{784}{28,125}$$

and find 5.3455 as the cube root. The calculator tells us 5.3485 is the cube root of 153, so the Babylonian formula was far more accurate for 153 than for 113. The explanation has to do with the difference between the A^3 and H terms. 49 is almost 77% of 64, but 28 is less than 23% of 125. The ratio between A^3 and H has a big effect on the accuracy.

PROBLEM 5.23   Use the Babylonian formula to find the cube root of 2000. Use your calculator to see how close the Babylonian answer is to the actual value.

Problem 5.24   Use the Babylonian formula to find the cube root of 65,432.

The precision of these square roots and cube roots was important to the Babylonians. There is one remarkable tablet in Yale University's collection that shows the square root of 2 expressed to 3 sexigesimal digits, which is about 9 of our decimal digits. The tablet shows $[1; 24, 51,10]_{60}$, which we interpret as $1 + \dfrac{24}{60} + \dfrac{51}{3600} + \dfrac{10}{216,000}$. You can confirm with your calculator is 1.414212963.

You can also verify with your calculator that the square root of 2 is almost exactly that, not differing until the sixth decimal digit.

Why did the Babylonian scribe who made this tablet need to be so precise? What if he had written the square root of 2 as $[1; 24, 51]_{60}$, leaving off the 216,000ths? Well, then he would have gotten 1.414166666 as his answer, and 1.414166666^2 is only 1.999867. All we can say is that they apparently cared about accuracy very much.

Of all the tablets, the most famous one is called Plimpton 322 (presumably because it was found by professor Plimpton of Columbia University after he had already found 321 others). This tablet shows that the Pythagorean theorem was well known and understood approximately 1000 years before Pythagoras was born. It lists whole numbers that could be the lengths of the sides of right triangles and the squares of these numbers.

The smallest right triangle with whole number sides is 3-4-5. The squares of those sides are 9-16-25, and of course $9 + 16 = 25$.

The Babylonians had a formula for finding an unlimited number of additional right triangles. The three sides were described as $\mathbf{p^2 - q^2}$, $\mathbf{2pq}$, and $\mathbf{p^2 + q^2}$, where p and q were whole numbers.

When $p = 2$ and $q = 1$, this formula gives

$$2^2 - 1^2 \quad (2)(2)(1) \quad 2^2 + 1^2$$
$$3 \qquad\qquad 4 \qquad\qquad 5$$

When $p = 3$ and $q = 2$, we get the 5-12-13 right triangle.

$$3^2 - 2^2 \quad (2)(3)(2) \quad 3^2 + 2^2$$
$$5 \qquad\qquad 12 \qquad\qquad 13$$

When $p = 3$ and $q = 1$, we get the 8-6-10 right triangle.

$$3^2 - 1^2 \quad (2)(3)(1) \quad 3^2 + 1^2$$
$$8 \qquad\qquad 6 \qquad\qquad 10$$

When $p = 4$ and $q = 1$, we get the 15-8-17 right triangle.

$$4^2 - 1^2 \quad (2)(4)(1) \quad 4^2 + 1^2$$
$$15 \qquad\qquad 8 \qquad\qquad 17$$

EXERCISE

Find the right triangle when p = 7 and q = 4.
Find the right triangle when p = 11 and q = 5.

EXERCISE

Use algebra to prove the Babylonian formula always works.

Suggestions for book or Internet research	
History, archeology	Mesopotamia Babylonia and Assyria Cuneiform Babylonian clay tablets
Religion, culture	Gilgamesh
Mathematical topics	Babylonian mathematics Mari numerals Babylonian multiplication Babylonian fractions Plimpton 322

ANSWERS TO PROBLEMS

5.1

This is a "3-digit" number since there are three groups of triangular wedges.

The group on the left is 15, because it has one wedge pointing to the side and 5 wedges pointing down.

The group in the middle is 2, because it has 2 wedges pointing down.

The group on the right is 37, because it has 3 wedges pointing to the side and 7 groups pointing down.

$$[15, 2, 37]_{60} \text{ therefore is} (15 \times 60^2) + (2 \times 60^1) + (37 \times 60^0),$$
$$\text{or } (15 \times 3600) + (2 \times 60) + (37 \times 1),$$
$$\text{or } (54{,}000) + (120) + (37)$$
$$\text{or } 54{,}157.$$

5.3

Converting our 8533 to Babylonian wedges requires figuring out the biggest power of 60 (1, 60, 3,600, 216,000, etc.) less than 8533. This is 3600. So, we divide 8533 by 3600, and we get 2 with a remainder of 1333. Our 3-digit number is $[2, x, x]_{60}$ with the x's yet undetermined.

We next divide 60 into 1333, and we get 22 with a remainder of 13. This allows us to complete the 3-digit number as $[2, 22, 13]_{60}$.

The Babylonians would write this as follows:

5.5

$$[55,30,19]_{60} + [25,34,49]_{60} = [1,21,5,8]_{60}$$

We start by lining the numbers up in columns and adding the last column:

$$
\begin{array}{rrr}
55 & 30 & 19 \\
25 & 34 & 49 \\
\hline
 & & 68
\end{array}
$$

68 is not a valid "digit" in base-60, so we carry the 60 (as a 1 in the next column) and leave the 8:

$$
\begin{array}{rrr}
 & 1 & \\
55 & 30 & 19 \\
25 & 34 & 49 \\
\hline
 & & 8
\end{array}
$$

Next we add the middle column:

$$
\begin{array}{rrr}
 & 1 & \\
55 & 30 & 19 \\
25 & 34 & 49 \\
\hline
 & 65 & 8
\end{array}
$$

65 is not a valid "digit" in base-60, so we carry the 60 (as a 1 in the next column) and leave the 5:

$$
\begin{array}{rrr}
1 & 1 & \\
55 & 30 & 19 \\
25 & 34 & 49 \\
\hline
 & 5 & 8
\end{array}
$$

Finally, we add the first column:

$$
\begin{array}{rrr}
1 & 1 & \\
55 & 30 & 19 \\
25 & 34 & 49 \\
\hline
81 & 5 & 8
\end{array}
$$

81 is not a valid "digit" in base-60, so we carry the 60 (putting a 1 in the next column) and leave the 21:

$$
\begin{array}{rrrr}
 & 1 & 1 & \\
 & 55 & 30 & 19 \\
 & 25 & 34 & 49 \\
\hline
1 & 21 & 5 & 8
\end{array}
$$

We can check our work by converting the base-60 numbers to base-10. We will see that we have just done the addition problem $199,819 + 92,089 = 291,908$.

5.7

$$[2,18,53,20]_{60} - [1,23,20,0]_{60} = [55,33,20]_{60}$$

We can line these numbers up, column by column, and start subtracting.

$$
\begin{array}{rrrr}
2 & 18 & 53 & 20 \\
- \quad 1 & 23 & 20 & 0 \\
\hline
 & & 33 & 20 \\
\end{array}
$$

20 minus 0 leaves 20 in the ones column.

53 minus 20 leaves 33 in the 60s column.

But 18 minus 23 (in the 3600s column) is impossible without borrowing, so we borrow from the 216,000s column. That reduces the 216,000s column from 2 to 1, and adds 60 to the 3,600s column:

$$
\begin{array}{rrrr}
1 & 78 & 53 & 20 \\
- \quad 1 & 23 & 20 & 0 \\
\hline
 & & 33 & 20 \\
\end{array}
$$

Now we can finish the problem:

$$
\begin{array}{rrrr}
1 & 78 & 53 & 20 \\
- \quad 1 & 23 & 20 & 0 \\
\hline
0 & 55 & 33 & 20 \\
\end{array}
$$

To check our work, we can convert the base-60 numbers to base-10, and we will see that we just performed the subtraction problem $500,000 - 300,000 = 200,000$.

5.9

N	$17 \times N$	N	$7 \times N$
1	17	6	1, 42
2	34	7	1, 59
3	51	8	2, 16
4	1, 8	9	2, 33
5	1, 25	10	2, 50
N		$7 \times N$	
20		5, 40	
30		8, 30	
40		11, 20	
50		14, 10	
60		17	

FIGURE 5.19 A table for multiplying by 17.

Notice that the table can be built step by step adding previous results. Once you enter 17 for $N=1$, you double 17 for $N=2$. You can add 17 and 34 to get the value for $N=3$. Then double the value for $N=2$ for $N=4$, and add the values for $N=2$ and $N=3$ to get the value for $N=5$, and so on. Again, the symbol 17 represents both the value for $N=1$ and $N=60$.

5.11

$$23^2 = 529.\ 8^2 = 64.\ 23+8 = 31.\ 31^2 = 961.$$
$$961 - 529 - 64 = 368.$$
$$368 / 2 = 184.$$

5.13

Since 10 divides evenly into 60, we convert $\dfrac{7}{10}$ to its equivalent, $\dfrac{42}{60}$.

$$\frac{7}{10} = [0; 42]_{60}.$$

Since 20 divides evenly into 60, we convert $\dfrac{3}{20}$ to its equivalent, $\dfrac{9}{60}$.

$$\frac{3}{20} = [0; 9]_{60}.$$

Since 8 does not divide evenly into 60, but does divide evenly into 3600, we convert $\dfrac{5}{8}$ to its equivalent, $\dfrac{2250}{3600}$. 2250 divided by 60 is 37 with a remainder of 30.

$$\frac{5}{8} = [0; 37, 50]_{60}.$$

We could have also gotten this result by taking the base-60 representation of $\dfrac{1}{8}$ and multiplying it by 5.

5.15

$$x^2 + 10x = 96.$$

The modern approach is to set one side of the equation equal to zero, factor, and say that the numbers that make the factors equal zero are the solutions for x.

$$x^2 + 10x - 96 = 0$$
$$(x+16)(x-6) = 0$$
$$(x = -16) \text{ or } (x = 6)$$

The Babylonian steps are as follows:

10 divided by 2 is 5.
5 squared is 25.
25 added to 96 is 121.
The square root of 121 is 11.
5 subtracted from 11 is 6.
6 is the value of x.

Observe that the Babylonians do not find the second solution, because it is a negative number.

5.17

For $x^2 + 2x = 399$, the Babylonian steps are as follows:

2 divided by 2 is 1.
1 squared is 1.
1 added to 399 is 400.
The square root of 400 is 20.
1 subtracted from 20 is 19.
19 is the value of x.

Again, there is a second solution the Babylonians do not find, -21.

5.19

The largest square less than 153 is 144, which is 12^2. So we express 153 as $12^2 + 9$ and find these three key numbers:

$$12,$$

$$\frac{9}{2 \cdot 12}, \text{ and}$$

$$\frac{9^2}{8 \cdot 12^3}, \text{ or}$$

$$12,$$

$$\frac{9}{24}, \text{ and}$$

$$\frac{81}{13,824}$$

and find 12.36914 as the square root. 12.36932 is what the calculator gives, rounded to 5 decimal places, so the Babylonian method is extremely accurate here.

5.21

The largest square less than 2000 is 1936, which is 44^2. So we substitute $44^2 + 64$ for 2000 and compute.

$$44,$$
$$\frac{64}{2 \cdot 44}, \text{ and}$$
$$\frac{64^2}{8 \cdot 44^3}, \text{ or}$$

$$44,$$
$$\frac{64}{88}, \text{ and}$$
$$\frac{4096}{681,472}$$

and find 44.7212622 as the square root.

What do we get if we square this? $(44.7212622)^2 = 1999.991293$

If we use the calculator to get the square root of 2000, we get 44.7213596

The Babylonian method is rather impressive.

5.23

The largest cube less than 2000 is 1728, which is 12^3. So we substitute $12^3 + 272$ for 2000 and compute.

$$12,$$
$$\frac{272}{3 \cdot 12^2}, \text{ and}$$
$$\frac{272^2}{9 \cdot 12^5}, \text{ or}$$

$$12,$$
$$\frac{272}{432}, \text{ and}$$
$$\frac{73,984}{2,239,488}$$

and find 12.5966 as the cube root.

When we use the calculator to get the cube root of 2000, we get 12.5992.

6

MATHEMATICAL ARCHEOLOGY

Two big questions arise very naturally when students start to study the history of math.

One question arises when the students reflect on how ancient writing is preserved and realize that a large amount of ancient math has been lost. They wonder: *Is what we know about Egyptian math and Babylonian math and Chinese math hopelessly incomplete?*

The second question arises when students reflect on how strange ancient math looks compared to today's math. They wonder: *How in the world did we figure these things out?*

THE INCOMPLETENESS OF THE ARCHEOLOGICAL RECORD

The problem of lost material is a real limitation. We think, for example, the ancient Chinese had a special interest in magic squares, because the books that have endured through the centuries have commentaries on magic squares. Is it possible that the Chinese had other big interests (which we just don't know about yet) and that magic squares were a minor concern? Yes, it is possible.

Similarly, we think the Egyptians had a special interest in unit fractions, because the papyri that have endured have problems with unit fractions. Is it possible that the Egyptians were thoroughly comfortable with ordinary fractions, but the luck of what has endured and what has not leads us to think they specialized particularly in unit fractions? Yes, it is possible.

An Introduction to the Early Development of Mathematics, First Edition. Michael K. J. Goodman.
© 2016 John Wiley & Sons, Inc. Published 2016 by John Wiley & Sons, Inc.

Our sense of how complete or incomplete our knowledge is has a statistical basis. At any moment, scholars have a finite amount of material from ancient Egypt or ancient China or ancient Babylonia. And they have a conceptual understanding of the mathematical sophistication of these civilizations. They know the range of mathematical interests and accomplishments.

Now suppose something new is unearthed, a new cache of papyri or inscriptions or cuneiform tablets. If the discovery gives us more examples of what we've already seen, then scholars grow more confident about the completeness of their view. But if the new discovery reveals something previously unseen, then scholars have to revise and expand their view. It turns out the situation has been relatively stable for a long time.

In the case of the Babylonians, we have tens of thousands of cuneiform tablets about mathematics. New ones are found and are deciphered all the time. Again and again, it is confirmed that the Babylonians represented numbers, did calculations, and solved equations the way we think they did. What might be new is a particular kind of problem, within the scope of the algebra, plane geometry, and arithmetic we already knew they knew. But it is inconceivable now that the Babylonians knew whole other areas of math, like calculus or fractal geometry, because we've never seen a tablet with calculus or fractal geometry on it.

Still, it makes sense to reflect on just how little we know about some civilizations, about how small our sample is. One historian of math summarized the scope of the Egyptian papyri by noting that we have only a handful of documents covering a period of several centuries. Let's put that in perspective by imagining a similar scenario regarding American mathematics:

Imagine what people 1000 years from now might think about math in America over the last few centuries if all they had were a few fragmentary records, like these:

- Billing data for a credit card company for one month in 2013
- Chapter 4 of a geometry textbook from the Civil War era
- George Washington's notes on a survey of 50 square miles of rural Virginia
- The sports section of a newspaper from one day in 1968

Our friends 1000 years in the future would certainly know a lot about our number system. The credit card data would show that we knew how to make and use large numbers; credit card companies like Visa and American Express use 16-digit numbers to identify their *customers' accounts*. The transactions, *purchases* and *payments*, would show the results of addition and subtraction, to two decimal places because we use dollars and cents. *Interest charges* on unpaid balances would show our ability to compute percentages. *Payment due dates* would show our expertise with calendars. *Addresses* would show house numbers and zip codes. It would be clear that ordinary citizens and ordinary businesses were quite familiar with numbers in a variety of roles.

Chapter 4 of that Civil War era geometry textbook might be about congruent triangles. Perhaps, our future friends would think that we were a society obsessed with proving that triangles were identical to each other if they had various combinations of equal sides and equal angles. There is a side-side-side theorem, an angle-side-angle theorem, a side-angle-side theorem, a side-side-angle theorem, and a hypotenuse-leg theorem. But the rest of geometry (more than 99% of it) would be undocumented.

George Washington's surveying notes might include a wealth of data about areas, distances, elevations, and directions. Some basic trigonometry would be there. Some units of measure would be there, like acres, miles, feet, and degrees.

And that 1968 sports section might have box scores from baseball, showing batting averages, earned run averages, innings pitched, and men left on base. If there is horse racing data, obscure distances like furlongs would be mentioned, as would betting odds expressed as ratios. A story about track and field might introduce metric measurements and sprinting times measured to hundredths of a second. Players' salaries, attendance figures in the thousands, and league standings would all feature numbers requiring interpretation.

But entirely missing (in this doomsday scenario of the record of American mathematics) would be thousands of academic journals and papers explaining research in number theory, combinatorics, computer code, calculus, set theory, statistics, fractals, and many other fields of mathematics. Entirely missing would be the practical applications of insurance, finance, and economics. How we taught mathematics would be unknown (except for that one chapter of that one textbook).

So if you imagine that these four things are the *only* fragments of mathematical information our future archeologists have access to, you will appreciate what it is like for modern scholars to try to understand the math of ancient civilizations.

At this point, we think we have a pretty complete record of the Babylonians and the Greeks, for example, but we know we have a mostly incomplete record of the Mayans and the Chinese.

THE STRANGENESS OF THE ANCIENT NUMBER SYSTEMS

Ancient people wrote in languages we don't easily read and represented numbers in ways unlike our own. Think about how we would decipher the aforementioned credit card data, geometry textbook, surveying notes, and sports reports if all the words were written in French and all the numbers were written as Roman numerals.

Let's put ourselves, briefly, in the shoes of a mathematical archeologist.

Suppose we came upon the somewhat broken item represented in Figure 6.1 in the course of an excavation.

It would be reasonable to wonder whether this was a bit of a story or a poem or part of a calendar or a mathematical table or some student's handwriting practice. At first glance it looks perplexing.

But answers don't come quickly in archeology. Lots of theories are tested and discarded. Comparisons are made to previous discoveries and interpretations. The context of a new find can push the analyst's thoughts in certain directions.

Recalling mathematical tables (like Egyptian tables of fractions and Babylonian tables of Pythagorean triples) might stimulate us to consider that this fragment might be a math table too. There is something curious about the first and second lines: the

			IP	**P**	**PI**	**PII**	**PIII**			
			IP	P	PI	PII	PIII	I{		
		PIII	{	{II	{IP	{PI	{PIII	{{		
	{II	{P	{PIII	{{I	{{IP	{{PII	{{{			
	{II	{PI	{{	{{IP	{{PIII	{{{II	{{{PI	{L		
	{	{P	{{	{{P	{{{	{{{P	{L	{LP	L	
PI	PI	{II	{PIII	{{IP	{{{	{{{PI	{LII	{LPIII	LIP	L{
PII	PII	{IP	{{I	{{PIII	{{{P				L{III	L{{
PIII	PIII	{PI	{{IP	{{{II	{L	LPI				L{{{
	I{	{PIII	{{PII	{{{PI	{LP					{Z
							{I			
	{{	{{{	{L	L				{Z		Z

FIGURE 6.1   A fragment an archeologist might find and try to interpret.

signs IP, P, PI, PII, and PIII are repeated in the same columns. The PIII at the end of line 1 (and near the end of line 2) shows up again at the start of line 3 and appears in several other places too, sometimes following a squiggle. The first and last columns show patterns characteristic of counting: in the first column, the I's increase from PI to PII to PIII; in the last column, the squiggles increase in several adjacent rows.

Let's return to the repetition of symbols in line 1 and line 2. If this is a counting table of some kind, we might reflect that adding zero to any number gives us the same number, and multiplying any number by 1 gives us the same number. This is reason enough to explore the idea that this is an addition table or a multiplication table. The first column has values (PI, PII, and PIII) that are also in the first row, and that too is characteristic of addition and multiplication tables. Look at the last row and the last column: similar combinations of squiggles and L's are in the same sequence, which is also characteristic of addition and multiplication tables. In fact, there is a symmetrical arrangement of identical symbols along the diagonal lines from the upper right down to the lower left, which is in accord with our knowledge in arithmetic that $(a + b) = (b + a)$ and $(a \times b) = (b \times a)$.

If you have convinced yourself that this is a student's multiplication table, you are correct. If trying to crack the code and decipher this fragment appeals to you, take a pause here before reading more. How the table was constructed will be explained next.

In Figure 6.2, there is a multiplication table, the kind that used to be printed in notebooks for elementary school children.

You can see that even though there is no text to tell us that numbers along the top are multiplied by numbers along the side, we can figure out easily that the numbers in the table are the results of multiplication. Reading the columns and rows shows us some familiar, regular patterns. If a number were omitted, or even wrong, we would know what to add or to change.

Now let's take that multiplication table and convert all the numbers to Roman numerals.

This "Roman numeral multiplication table" shown in Figure 6.3 is a direct translation of the previous multiplication table. The only difference that might make us pause for a moment is that some numbers are broken into two parts and written as

	1	2	3	4	5	6	7	8	9	10
1	1	2	3	4	5	6	7	8	9	10
2	2	4	6	8	10	12	14	16	18	20
3	3	6	9	12	15	18	21	24	27	30
4	4	8	12	16	20	24	28	32	36	40
5	5	10	15	20	25	30	35	40	45	50
6	6	12	18	24	30	36	42	48	54	60
7	7	14	21	28	35	42	49	56	63	70
8	8	16	24	32	40	48	56	64	72	80
9	9	18	27	36	45	54	63	72	81	90
10	10	20	30	20	50	60	70	80	90	100

FIGURE 6.2 A multiplication table. Before calculators became ubiquitous, schoolchildren memorized these.

	I	II	III	IV	V	VI	VII	VIII	IX	X
I	I	II	III	IV	V	VI	VII	VIII	IX	X
II	II	IV	VI	VIII	X	XII	XIV	XVI	XVIII	XX
III	III	VI	IX	XII	XV	XVIII	XXI	XXIV	XXVII	XXX
IV	IV	VII I	XII	XVI	XX	XXIV	XX VIII	XXX II	XXX VI	XL
V	V	X	XV	XX	XXV	XXX	XXX V	XL	XLV	L
VI	VI	XII	XVII I	XXIV	XXX	XXX VI	XLII	XL VIII	LIV	LX
VII	VII	XI V	XXI	XX VIII	XXX V	XLII	XLIX	LVI	LXIII	LXX
VII I	VII I	XV I	XXI V	XXXI I	XL	XLVII I	LVI	LXIV	LXXII	LXX X
IX	IX	XV III	XX VII	XXX VI	XLV	LIV	LXIII	LXX II	LXXX I	XC
X	X	XX	XXX	XL	L	LX	LXX	LXX X	XC	C

FIGURE 6.3 The multiplication table with Roman numerals substituted for the Arabic numbers we ordinarily use.

one part on top of the other. For example, the Roman 35, XXXV, appears as XXX over V in the intersection of the grid for multiplying V by VII (5 by 7). This apparently allowed the person who made this multiplication table to keep the width of the columns fairly constant.

Imagine seeing Figure 6.3 without having previously seen Figure 6.2. You would probably be able to figure out that Figure 6.3 was a multiplication table anyway, since you know Roman numerals, and since you have seen mathematical tables before.

But we can make the task harder by leaving out parts of the table.

Figure 6.4 simulates what happens in the real life of an archeologist—instead of getting a whole artifact, the archeologist gets pieces. Weathering, breakage, and even theft reduce many important items to mere fragments, which then have to be

			IV	V	VI	VII	VIII		
			IV	V	VI	VII	VIII	IX	
			VIII	X	XII	XIV	XVI	XVIII	XX
			XII	XV	XVIII	XXI	XXIV	XXVII	XXX
		XII	XVI	XX	XXIV	XXVIII	XXXII	XXXVI	XL
	X	XV	XX	XXV	XXX	XXXV	XL	XLV	L
VI	VI	XII	XVIII	XXIV	XXX	XXXVI	XLII		LIV	LX
VII	VII	XIV	XXI	XXVIII	XXXV			LVI	LXIII	LXX
VIII	VIII	XVI	XXIV	XXXII	XL					LXXX
	IX	XVIII	XXVII	XXXVI	XLV				X I	XC
	X	XX	XXX	XL	L				XC	C

FIGURE 6.4   The Roman numeral multiplication table has been damaged.

			IP	P	PI	PII	PIII		
			IP	P	PI	PII	PIII	I{	
			PIII	{	{II	{IP	{PI	{PIII	{{
			{II	{P	{PIII	{{I	{{IP	{{PII	{{{
		{II	{PI	{{	{{IP	{{PIII	{{{II	{{{PI	{L
	{	{P	{{	{{P	{{{	{{{P	{L	{LP	L
PI	PI	{II	{PIII	{{IP	{{{	{{{PI	{LII	{LPIII	LIP	L{
PII	PII	{IP	{{I	{{PIII	{{{P	{LII	{LI{	LPI	L{III	L{{
PIII	PIII	{PI	{{IP	{{{II	{L	{LPIII	LPI	L{IP	L{{II	L{{{
	I{	{PIII	{{PII	{{{PI	{LP	LIP	L{III	L{{II	L{{{I	{Z
	{	{{	{{{	{L	L	L{	L{{	L{{{	{Z	Z

FIGURE 6.5   The Roman V, X, and C have been replaced by less familiar symbols.

reconstructed. If all the pieces are available, it's like assembling a jigsaw puzzle; if some pieces are missing, it's even more difficult (and guesses have to be made for the missing parts).

And we can make this particular puzzle harder by replacing some of the familiar Roman numeral symbols with substitutes. In Figure 6.5, the familiar V for 5 has been replaced by a P; the familiar X for 10 has been replaced by a squiggle; and the familiar C for 100 has been replaced by a Z.

This is the table as you first saw it, several pages ago. Knowing what it is and how it was constructed makes it appear obvious, but archeologists start out without knowing those things. Figure 6.1 was a curiosity, something we didn't immediately recognize and understand. Figure 6.5, which is exactly the same as Figure 6.1, is something we immediately understand.

Archeologists are like code-breakers. When they come across an unknown number system, they have to be patient, logical, and creative. Their biggest clues often come from the meaning of the words associated with the unknown numbers. Ancient people may have written about the 5 planets they could see in the sky or the 4 seasons of the calendar year, pinpointing their symbols for 5 and 4. If they worshipped one god or a number of gods, the symbol for 1 or the symbol for their number of gods might be easy to find.

It makes sense to guess that longer strings of symbols represent bigger numbers. (The Roman 10 is X, and the Roman 18 is XVIII.) It makes sense that symbols that occur more frequently represent more common numbers. (Numbers like 1, 2, 3, and 4 occur much more frequently in ordinary human affairs than numbers like 31, 87, and 205.) There may also be some internal logic among the symbols. (The top half of the Roman 10, X, looks like the Roman 5, V.)

The context of the numbers also provides clues. Ancient kings would boast about ruling large numbers of people and owning large numbers of things, not small numbers of people and things. Ancient travelers would need directions to a place 500 miles away but not to a place 5 miles away. The context might suggest that a calculation is being made or an equation is being solved.

And finally, an archeologist could rely on analogies to number systems that have already been deciphered. The additive, multiplicative, positional, and subtractive features of Egyptian, Chinese, Babylonian, and Roman numbers might occur in other systems.

Working against the archeologist are three key factors. First, he might have only a piece of an artifact. Second, he is relying on the accuracy of the scribe who created the artifact: What if the scribe made or copied a mistake? And third, there is ambiguity among many ancient symbols. Just as our 9 can mean exactly 9, or 9000 in a number like 9242, the single Babylonian triangular wedge could mean 1 or 60 or 3600, and a capital M in Ionic Greek might be 40, but M is also a signal to make a very large number by multiplying by 10,000. Deciphering ancient writing and numbers sometimes takes decades.

Deciphering Roman numerals never posed a problem, because even though they are 2000 years old, we never lost contact with them. They didn't have to be rediscovered and interpreted.

The Roman Empire dominated the Mediterranean world for centuries, and if you wanted to study Roman architecture, Roman engineering, Roman law, or Roman literature, you would have no trouble finding material. But if you wanted to study Roman mathematics, you'd soon be scratching your head—there appears to be nothing to find, except for their quaint alphabetical system of numerals: I, V, X, L, C, D, and M for 1, 5, 10, 50, 100, 500, and 1000. We know the names of dozens of Greek mathematicians who came before the Romans, and we know the names of Arabian and European mathematicians who came after, so why don't we know the names of Roman mathematicians?

Because there are none. The Romans were interested in practical applications of math: how to build things (roads, houses, stadiums, temples, cities), and how to organize things (armies, governments), and how to make money (trade, taxation). They were not interested in the theoretical aspects of math that so intrigued the Greeks.

It is sometimes said that Roman numerals themselves held back progress in math in Europe for hundreds of years, but that seems to be an exaggerated complaint.

Multiplying CCCXLIII by VII to get MMCDI certainly looks complicated and clumsy, but surely you don't rule the world for 500 years if you can't do arithmetic. The Romans had their methods, and the children of aristocrats learned practical math as a routine part of their education.

Our interest in Roman mathematics is almost exclusively an interest in their numerals. Their system is like the Greek systems (which we will describe in the next chapter) in that letters of the alphabet represent digits and repeating the symbols has a multiplying effect (which is also like the Egyptian system). X is 10, while XX is 20, and XXX is 30. Lumping by 5s (V instead of IIIII, L instead of XXXXX) is like the Attic Greek Γ (penta), but a feature that appears to be original with the Romans is the subtracting effect of placing a smaller number first: IX is 9 because it is 1 before 10, XL is 40 because it is 10 before 50.

The subtractive element can occur before the base-10 number or the lump of 5. So, IV is 4, and XC is 90; CD is 400, and CM is 900.

Additive elements follow the lump of 5: VII is 7 because there are 2 after the 5.

Additive elements cascade: LXVII is 67, because there is a 7 after the 10 after the 50.

To make large numbers, larger than a few thousand, a bar is placed over the part that is multiplied by 1000.

$\overline{\text{XXVI}}$DCCXIX is 26,719 since XXVI has a bar over it.

It will probably be fun for you, and a little nostalgic, to try to write some of our numbers as Roman numerals and to interpret or translate some Roman numerals into our numbers.

$$DLV = 500 + 50 + 5 = 555$$

An archeologist might take special notice of the fact that three different symbols here all represent "five" of something: 5 hundreds, 5 tens, or 5 ones. He might observe that the number is read left-to-right, and he might observe that a quantity that is a 3-digit number to us is also a 3-character number in this system.

$$MMMCCCXXXIII = 3000 + 300 + 30 + 3 = 3333$$

An archeologist might notice here the repetition of symbols and also be struck that what is a 4-digit number to us requires 12 characters in Roman numerals. The sequence of symbols for larger quantities coming first is also illustrated: the thousands come before the hundreds, which come before the tens and the ones.

$$\overline{\text{MCMIV}}\text{DCCCLXXIV} = \left[1000 \cdot (1000 + 900 + 4)\right] + (500 + 300 + 50 + 20 + 4)$$
$$= 1,904,874$$

The sequence observed for larger quantities before smaller quantities is totally out of whack in this number. The symbols represent (left-to-right) the following: 1000, 100, 1000, 1, 5, 500, 100, 100, 100, 50, 10, 10, 1, 5. The archeologist would not throw his hands up in frustration that his understanding of the previous pattern was wrong. He would use it as a framework to interpret the additional rules of this system, and thus he would start to understand the subtractive element (100 before 1000 = 900) and the significance of the bar over the first five symbols (multiply that part by 1000).

Exactly how the Romans chose to use this system is unclear. Certainly, it has some similarity to the Greek systems, but also there is a likely connection to the Etruscans, who ruled central Italy before the Romans conquered them. The region north of Rome, Tuscany, was called Etruria in ancient times, and a good deal of Etruscan culture, possibly including their numerals, was absorbed into Roman culture.

While the Roman system poses no difficulties for mathematical archeologists, the Mayan system was an utter enigma to the Europeans who first tried to decipher it. We know almost nothing about the procedures and calculation methods of Mayan mathematics. And we can blame the Spanish conquistadores for this: in their zeal to convert the Mayans to Christianity, they destroyed almost every written Mayan record.

What survives is interesting for several reasons. First, we can say (for obvious and incontrovertible reasons) that Mayan mathematics developed with no help from the Egyptians, Babylonians, Chinese, and others, and the geographic isolation of the Mayans (in southern Mexico and northern Central America) thus makes it clear that mathematics was "invented" multiple times, and that any human group, with the motivation to do so, could develop complex mathematical ideas.

A second point of interest is the Mayan number system itself. It is a base-20 system, and we ourselves, and many ancient people preferred a base 10-system. It will be useful for the student to learn the Mayan numerals and practice converting their numbers to ours and ours to theirs. The conversions will sharpen arithmetical skills and will also give some insight about the task of the mathematical archeologist when faced with a new number system to decipher and understand.

Third, from the fragments we have, we know that the Mayans were fantastic observers and calculators. They were spectacularly accurate in determining the length of the solar year and the recurrence of astronomical phenomena, like the positions of planets.

Fourth, unlike a number of ancient cultures, the Mayans used a symbol for zero. This very important mathematical value had to be inferred in some cultures, and the possibility for error was significant.

Let's look closely at the Mayan representation of a number.

We read the stack of dots and lines in Figure 6.6 as four "digits" where digits can range between 1 and 19. The top digit, three dots over a line, represents 8. Horizontal lines represent 5. Dots count as 1 each. This is 8, but 8 what? There is some controversy, but the simple answer is 8×20^3, which is 64,000.

The next digit is 16 (three lines, each worth 5, and one dot). And this is 16×20^2, or 6400. The third digit (we read down) is 9. This is 9×20, or 180.

The last digit is 3, and this doesn't have to be multiplied by anything. So the full number is $64,000 + 6,400 + 180 + 3$, or 70,583.

Just as we multiply digits by powers of 10, the Mayans multiplied them by powers of 20. The controversy alluded to above is that in some Mayan writing, the scaling factor for one of the 20s is only 18. Strictly multiplying by 20s, the digits represent multiples of 1, 20, 400, 8000, 160000, 3200000, and so on. Some calculations look more sensible if we assume the digits represent multiples of 1, 20, 360, 7200, 144000, 2880000, and so on. This wrinkle in the system is assumed to be a convenience for their calendar, as 360 is so close to the number of days in a year, and substituting (18×20) for (20×20) one time gives the revised list of place values.

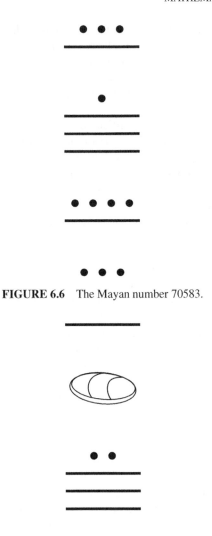

FIGURE 6.6   The Mayan number 70583.

FIGURE 6.7   The Mayan number 40344.

In a vertical column of digits, like the column representing 70,583, there is no need to employ the Mayan symbol for zero, because 70,583 is a combination of a certain non-zero number of 8000s and 400s and 20s and ones. However 40,344 is, when we break it down, $(5 \cdot 8000) + (0 \cdot 400) + (17 \cdot 20) + (4 \cdot 1)$, so the Mayans represented 40,344 as the dots and lines and the special symbol for zero shown in Figure 6.7. It's not so clear what the symbol for zero is supposed to represent; some people say it's a snail's shell.

For our purposes, let's forget that wrinkle of 18 taking the place of 20 in one digit and see how we would convert one of our numbers into a Mayan number. Let's take a large number: 345,678. The starting consideration is always the powers

FIGURE 6.8 The Mayan number 345678.

of 20...1, 20, 400, 8000, 160000.... The largest of these that is smaller than 345,678 is 160,000, so we divide 345,678 by 160,000. It goes in twice, with a remainder of 25,678.

This makes the first "digit" of our Mayan number 2.

Next we divide 25,678 by 8000. 8000 goes in 3 times, with a remainder of 1678. This makes the second "digit" of our Mayan number 3.

Next we divide 1678 by 400. 400 goes in 4 times, with a remainder of 78. This makes the third "digit" of our Mayan number 4.

Next we divide 78 by 20. 20 goes in 3 times, with a remainder of 18. This makes the fourth "digit" of our Mayan number 3 and the final digit 18.

This process, which seems so straightforward, is easy only because we have already cracked the code and learned what the dots and lines mean. The key was translating the codexes that the conquistadores didn't burn. A codex looks like a very long comic strip, folded up like an accordion. They were made from plant fibers pressed into a strong paper and illustrated with bright pigments. The texts are generally historical records celebrating the careers of Mayan rulers, and the numerical parts show what historical records typically show: how long a king reigned, how many victories he won, and so on.

In the remainder of this chapter, you get to play the role of a mathematical archeologist encountering an unknown number system.

Pretend that the following story is the translation of an ancient text, where the experts are able to figure out the words confidently but are less sure about the numbers, which are represented by symbols. Use the logic of the story and analogies to actual ancient number systems to decipher the numbers in the story. (For convenience, the story is separated into 5 paragraphs.)

1. So we went on the journey, to pay tribute to King Mike. The road was difficult, and the trip to his city, which often takes only Թ days, this time took Ժ Գ. All together, Ի Դ of us set out, but Գ were killed by bandits who attacked us, and Բ more became ill and died, so only Ժ Թ of us arrived at King Mike's city.
2. The distance is great. I think it is more than Մ miles. Sandro says it is less. He measured by counting steps, and he said we covered Ժ Ձ miles on day Ա and Ժ Ե miles on day Բ, but with the terrible weather and the bandits' attack, he stopped counting, so the true distance is unknown.
3. We brought considerable gifts with us, to flatter King Mike. We had Ժ Ա fine gems. I had contributed Դ, and Baldini had also contributed Դ, and Sandro had contributed Գ. We also had Ի Ա carved wooden statues of King Mike. King Mike is somewhat on the vain side, and he likes to have carved wooden statues of himself around. Է of our artists each made Գ statues. It doesn't take long to make a passable statue—King Mike's eyesight isn't so good.
4. When we arrived at the city (after Ժ Գ days), it was a magnificent sight. There must have been at least Ձ houses, and at least S people in the city. We were impressed. They showed us a game that is popular among them. On a square board, with Ձ rows and Ձ columns, they place Լ Ե coins, leaving Ա square empty. Then they move the coins around, by some rules we could not quite understand, but there is shouting and laughter and gambling as the coins are moved.
5. King Mike thanked us for our gifts and sent us home with gifts also. He asked each of us to extend our hands, and he put a silver bar in every palm, Լ Ը in all. Then his henchman told him that Ե of our men had died on the journey, and King Mike gave me Ժ more silver bars to bring to their widows.

- - - - - *what follows is a discussion of how you might try to decipher the mathematical symbols in this story. You may want to stop here and try to be an archeologist before reading further* - - - - -

In paragraph [1], there are several clues. Ժ Գ must be a bigger number than Թ because the context suggests the difficult conditions made the journey last longer. Also, we can do some arithmetic with the number of people who set out and the number of people who completed the journey. It seems fair to conclude that

Գ + Բ + Ժ Թ = Ի Դ

This says the number who died on the way, plus the number who arrived, equals the number who set out.

The symbol Գ appears alone in the equation, but paired with Ժ in the number of days.

The symbol Թ appears alone as a number of days, but paired with Ժ in the equation. These facts tell us that symbols can be used meaningfully alone or in combinations, and they suggest that numbers expressed by two symbols are larger than numbers expressed by one symbol.

There is a huge context clue in paragraph [2]. Sandro started to calculate the true distance by counting steps, but stopped, so it makes sense that he counted on the first day and the second day. That suggests U is 1 and Բ is 2. We also see two symbols that both look like a capital U, Ս, and Ս, and we have to be careful not to confuse them. (They could be completely different numbers, or they could be two slightly different representations of the same symbol, like a handwriting variation.)

In paragraph [2], we see the symbol ժ again used in combinations, and we begin to strongly suspect it is a very common symbol for combinations. However, something surprising occurs regarding the number of symbols: if Ս and Ս are really different numbers, the one symbol number Ս seems to be a large number, from its context, contradicting the suggestion above that larger numbers involve more symbols.

Also, Բ appeared in paragraph [1] also—if it means 2, then 2 people died on the journey from illness.

Paragraph [3] has some hints about other one symbol numbers. It looks like we can make an "addition" equation with the number of gems and a "multiplication" equation with the number of artists and statues. We also see that it is possible to have a two symbol number that does not include the symbol ժ.

Paragraph [4] introduces some symbols that we don't see anywhere else, Ջ and S. All we can say is that the context makes them seem like large numbers, resolving the little mystery about whether one symbol numbers had to be smaller than two symbol numbers. In paragraph [1], there was a suggestion. In paragraph [2], there was an apparent contradiction of that suggestion, and now paragraph [4] completely undermines the suggestion. The gambling game gives us more useful arithmetic (for the size of the board) and additional confirmation that Ս probably is 1.

Paragraph [5] tells us that Ե must be the sum of Գ+Բ, the men who died, and if every man extends two hands to King Mike, then we have clues about numbers of silver bars being twice as large as numbers of men.

You might be able to translate virtually the whole story now, except for knowing the exact value of the larger numbers that are mentioned only once. I will tell you that the set of symbols used in this story is a real numeral system! It is an old Armenian system, and you can confirm your guesses via the Internet or a book. You are on the right track if you determined that 24 men set out to see King Mike, 19 arrived 13 days later, bringing 11 gems and 21 statues, and that they came home with 38 silver bars for themselves and 10 silver bars for the widows.

The Armenian numeral system turns out to have a lot in common with one of the Greek numeral systems, the one known as Ionic Greek, where letters of the Greek alphabet represent 27 distinct numbers. The Armenian alphabet has 36 letters that represent 36 distinct numbers.

7

CLASSICAL GREEK MATHEMATICS

You already know a great deal more about classical Greek mathematics than you think you do. Virtually, everything you learned in school about basic arithmetic, algebra, geometry, and trigonometry is classical Greek mathematics. Other groups added to the core of Greek mathematical knowledge—including, first, the Arabs who preserved it while Europe was in the Dark Ages, and then the Italians of the Renaissance, and finally and most significantly the French of the Enlightenment.

What we know about Greek mathematics—specific problems, specific people, books, methods, calculations, theories—dwarfs what we know collectively about the math of all other ancient civilizations combined. Greek work has been translated so exhaustively that we know the step-by-step procedure Archimedes used to work out the formula for the volume of a sphere and the step-by-step procedure Euclid used to prove that the supply of prime numbers was infinite.

And there should be no surprise about this classical Greek mathematical heritage. We already look back to the Greeks for the roots of our political organization and democracy. We look back to the Greeks for our concepts about beauty in art, about the structure of poetry and drama, and about the methods and concerns of philosophy. We are, to some extent, in awe when we consider the Greek contributions to our modern world.

It is a question worth asking: *why were the Greeks able to accomplish so much?* Confining the question only to mathematics, we can find half a dozen explanations.

First, there is the obvious answer of chronology: the Greeks came after the Babylonians, the Egyptians, and others, and they were thus able to learn from and build on the mathematical work of others.

An Introduction to the Early Development of Mathematics, First Edition. Michael K. J. Goodman.
© 2016 John Wiley & Sons, Inc. Published 2016 by John Wiley & Sons, Inc.

Second, geography favored the Greeks. Mainland Greece, the Greek islands in the Aegean Sea, and the Greek colonies around the eastern Mediterranean were situated at or near the major trade routes of antiquity. Just as goods moved from place to place, ideas moved from place to place, and a good mathematical idea discovered in Arabia or Syria or Turkey or Sicily or Egypt or Crete was known everywhere else in a relatively short amount of time.

Third, the Greek political system fostered more general education. In older societies, including the Egyptian and Babylonian (which we looked at) and groups contemporary to the Greeks (like the Persians), government was in the hands of an aristocratic minority and therefore education was reserved for the ruling class. In democratic Greek cities, where any citizen might be called upon to exercise judgment in civic and administrative matters, it was in the interest of the state to have a large pool of talented, educated men. Something like a Babylonian scribe school would have been an anachronism to the Greeks.

Fourth, the Greeks were wealthy. They could afford to sit around and discuss and theorize about algebra and geometry. In older societies, math was almost totally a practical science, a body of knowledge directed toward building city walls and streets and canals and irrigation ditches, figuring out the calendar, and collecting the taxes and managing the municipal payroll, and regulating day-to-day commercial, agricultural, and military transactions. Ancient Egyptians and Babylonians didn't have time to think about infinity or worry about prime numbers. Ancient Greeks did.

Fifth, the Greeks had a reverence for talent, ability, and accomplishment, encouraged by many of their leaders and especially by Alexander the Great. Alexander contributed deeply to mathematical development in several ways. By conquering all the parts of the known world thought to be worth conquering, he created a unified, stable society where people and ideas could move quickly. He appointed and promoted the best people. And he encouraged the expansion of great libraries, the greatest of all being the one in Alexandria, Egypt, where the knowledge of his entire empire was copied and stored and made available to the scholars of the time.

And sixth, competition among the Greek cities spurred further intellectual exploration within each one. If a genius in one city invented something or figured something out that gave his city an economic or military advantage, it was imperative for other cities to have their geniuses busily working to close the gap.

Competition among the cities also accounts for the fact that there are several Greek number systems. Two prominent systems are Attic Greek numerals and Ionic Greek numerals. Attic Greek is named for Attica, the region around Athens. A typical Attic Greek number is **X X H Δ Δ Δ Γ I I**, which is 2137.

The two **X**'s each represent 1000. **X** (chi) is the first letter of the Greek word for 1000.

H (eta) represents 100, and similarly it is the first letter of the Greek word for 100.

Δ (delta) represents 10. Observe that there are 3 of these in the number (making 30).

Γ (an old version of the Greek equivalent of "p" for "penta," even though it looks like the modern Greek letter gamma) represents 5.

I (iota) represents 1.

The symbols are lined up and added.

Ionic Greek is named not for the Ionian Sea, the sea between Greece and southern Italy, but for Ionia, a region of coastal western Turkey that was colonized by the ancient Greeks. Ionic Greek numerals also used the Greek alphabet, but in a quite different way. The number 2137 in Ionic Greek is \β ρ λ ζ.

The β (beta) ordinarily represents the number 2 (it is the second letter of the alphabet), but with the slash in front of it here, it becomes 2000. The effect of the slash is to multiply the number that comes after it by 1000. The ρ (rho) is a symbol for 100. The λ (lambda) is 30, and the ζ (zeta) is 7. In the Ionic Greek scheme, there are separate symbols for the numbers 1–9, the multiples of 10 from 10 to 90, and the multiples of 100 from 100 to 900. This requires 27 separate symbols, and the Greek alphabet has 24 letters that account for 24 of the 27. The extra three symbols are represented by older letters no longer in use after the alphabet was standardized with the current 24 letters.

There are many variations and complications for these numeral systems, and students will find them interesting or challenging or frustrating, depending on several factors.

\β ρ λ ζ could also be written \Β Ρ Λ Ζ (using the capital Greek letters), for example. Another complication is that throwing a big **M** (mu, ordinarily the symbol for 40) in front of (or behind, or sometimes even under) a string of numbers multiplies the string by 10,000.

It's the Greek mathematicians who really capture our attention and admiration, and there is a spectacular parade of them. There is no way to talk about them all in sufficient detail in a small book like this, and the selected highlights that follow are a very idiosyncratic (and roughly chronological) sampling of the author's favorites. The reader is encouraged to pursue any interesting topics further and is assured that there is a mountain of material to be found.

The first Greek mathematician of note is **Thales**, who was active around 600 BC. He came up with ideas that are essential building blocks of the geometry you learned in high school. For example, he pointed out the fact that vertical angles are equal: vertical angles are formed when two lines intersect. He pointed out that if you have a triangle and draw a straight line through the triangle that is parallel to one of the sides of the triangle, you form a similar triangle: similar triangles have exactly equal angles, and the ratio of the lengths of a pair of their corresponding sides is the same as the ratio of the lengths of any other pair.

In Figure 7.1, angles 1 and 3 are equal to each other, and angles 2 and 4 are equal to each other. Our eyes tell us this is true, but an ironclad rule of geometry also tells us: the angles that together form a straight line always add up to 180°. The Greeks would have said "two right angles" instead of 180° because the terminology at the time favored saying "two right angles," but of course it means the same thing.

Angles 1 and 2 form a straight line. Angles 2 and 3 form a straight line. Angles 3 and 4 form a straight line. Angles 4 and 1 form a straight line. Thales observed that you could add either angle 1 or angle 3 to angle 2 to make 180°, which meant angle 1 and angle 3 had to be the same size.

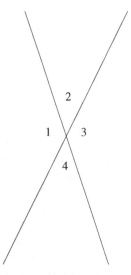

FIGURE 7.1   Thales asserted that intersecting straight lines formed pairs of equal angles.

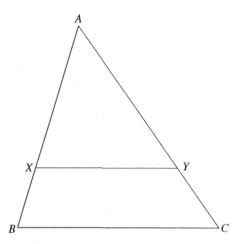

FIGURE 7.2   Thales asserted that the lengths of the sides of similar triangles made consistent ratios.

In Figure 7.2, $\triangle ABC$ and $\triangle AXY$ are similar, with BC parallel to XY. Now there are a host of equal ratios:

$$\frac{AX}{AB} = \frac{AY}{AC} = \frac{XY}{BC}$$

But also $\dfrac{AX}{AY} = \dfrac{AB}{AC}$, and $\dfrac{AX}{XY} = \dfrac{AB}{BC}$, and $\dfrac{AY}{XY} = \dfrac{AC}{BC}$

Even $\dfrac{BX}{CY} = \dfrac{AB}{AC}$ and $\dfrac{BX}{AB} = \dfrac{CY}{AC}$

A wonderful theorem attributed to Thales is that if you inscribe a triangle inside a circle and the longest side of that triangle is also the diameter of the circle, then the triangle is a right triangle. In Figure 7.3, AC is a diameter of the circle and B could be located at any point along the circumference above AC.

We can approach this several ways, but for simplicity, let's examine an equilateral triangle inscribed in a circle and a square inscribed in a circle.

In Figure 7.4, the 60° angles of the equilateral triangle intercept 120° arcs of the circle. (We use the phrase "intercept an arc" to indicate how much of the circumference of a circle is spanned when we extend the two sides of the angle until they meet the circle.) Observe that the three equal angles of the equilateral triangle appear to intercept three equal arcs of the circumference. Observe also that the ratio of any angle to its intercepted arc (60° : 120°) is 1 : 2. Since all three angles in the triangle are equal, it makes sense that all three arcs of the circle are equal.

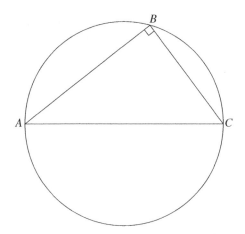

FIGURE 7.3 Thales asserted that a triangle inscribed in a semicircle had to be a right triangle.

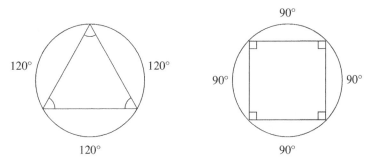

FIGURE 7.4 An equilateral triangle inscribed in a circle, and a square inscribed in a circle.

The corners of the square in Figure 7.4 meet the circle at four points, making four equal arcs on the circumference, and each of those arcs is therefore 90°. But each 90° angle of the square intercepts two of the four arcs, so each 90° angle intercepts a 180° arc. Here too, the ratio of the angles to their intercepted arcs (90° : 180°) is 1 : 2.

It looks like angles drawn inside a circle, on its circumference, intercept arcs that have twice as many degrees. We have just observed it, casually and empirically. Thales proved it rigorously. Back in Figure 7.3 then, we have a diameter as one side of the triangle. The arc of the circle below the diameter must be 180°, and the angle that intercepts it, angle ABC, must be 90°.

Pythagoras comes next and, ironically, the thing he is known best for is something we know for certain he didn't invent: $a^2 + b^2 = c^2$ for a right triangle. He may have presented it better than anyone, and his proof of it may be exceptionally clever (hundreds of different proofs are known), and he may have put it to use very creatively, but he certainly wasn't the first to discover it. It was known in Babylonia, long before Pythagoras lived, and it was known in China.

Pythagoras did, however, do a lot of other things with numbers. He was a pioneer in investigating properties of numbers and was the leader of a cult-like religion of numbers. Some of the religious mumbo-jumbo is amusing (e.g., the number 2 is associated with *opinion*, and the number 4 with *justice*), but Pythagoras' technical analysis of whole numbers is mathematically sound and interesting. For example, Pythagoras observed that any sequence of odd numbers, starting with 1, adds up to a perfect square:

$$1 + 3 + 5 + 7 + 9 + 11 = 36 = 6^2$$

Pythagoras worked on triangular numbers, those that are the sums of the first 2 numbers, 3 numbers, 4 numbers, and generally n numbers, which can be represented as triangular arrays of dots. The first six triangular numbers, shown in Figure 7.5, are 1, 3, 6, 10, 15, and 21.

Pythagoras observed that any square number is the sum of two consecutive triangular numbers. The identity $36 = 15 + 21$ is shown in Figure 7.6, and in any square array of dots a diagonal line could separate two triangular numbers.

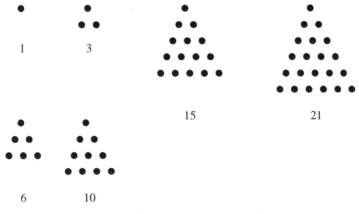

FIGURE 7.5   Triangular numbers.

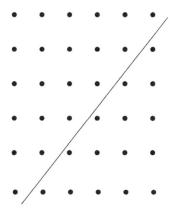

FIGURE 7.6   The sum of two consecutive triangular numbers is a square.

PROBLEM 7.1   Express 400 (20²) as the sum of two triangular numbers.

PROBLEM 7.2   Express 841 (29²) as the sum of two triangular numbers.

Pythagoras led a large group of mathematically talented people, so historically it might be inaccurate to give Pythagoras credit for something that someone else in his group came up with, but he was the philosophical and inspirational leader, so it seems reasonable to think that he in some way contributed to almost all the ideas credited to him. One of these ideas concerns the number of degrees in all the angles of a polygon. Pythagoras saw that polygons could be partitioned into triangles, and since the angles of each triangle summed to 180°, the angles of polygons summed to multiples of 180°. Quadrilaterals (four-sided figures like squares, rectangles, trapezoids, and parallelograms) summed to 360°. Pentagons to 540°. Hexagons to 720°. Figure 7.7 shows an octagon split into six triangles by lines drawn from one vertex to all nonadjacent vertexes. The degrees in the eight angles of the octagon sum to 6×180, or 1080°.

It was not much of a stretch to go from this result to a fairly surprising (but still fundamental) one: that the sums of the exterior angles of any polygon add up to 360°.

Exterior angles are formed by extending the sides of a polygon, uniformly either clockwise or counterclockwise, all the way around, as shown in Figure 7.8.

EXERCISE

Show that angles 1, 2, 3, 4, 5, 6, 7, and 8 in figure 7.8 must add up to 360°.

Another remarkable Pythagorean find is that the numbers that divide evenly into 220 add up to 284, and the numbers that divide evenly into 284 add up to 220. This is a rarity. Take an ordinary number, say 50. The numbers that divide evenly into 50 are 1, 2, 5, 10, and 25. These add up to 43. But the numbers that divide evenly into 43 do no add up to 50. In fact, the only number that divides evenly into 43 is 1. The numbers that divide evenly into 36 are: 1, 2, 3, 4, 6, 9, 12, and 18. They add up to 55. But the numbers that divide evenly into 55 do not add up to 36. (They are 1, 5, and 11, and they add up to 17.)

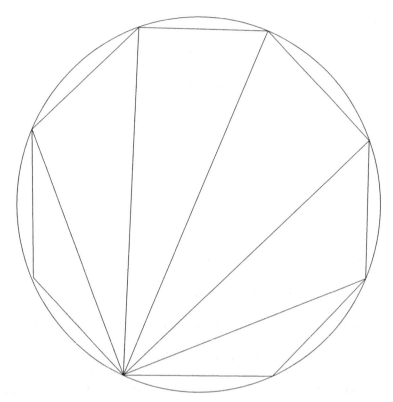

FIGURE 7.7 Pythagoras is credited with the observation that the sum of all the angles in a polygon is a multiple of the angles in a triangle.

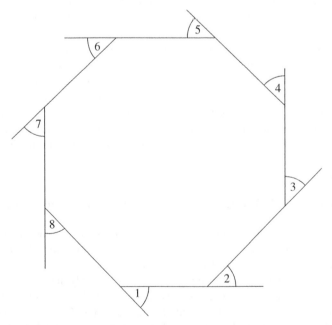

FIGURE 7.8 Pythagoras is credited with the observation that the sum of all the exterior angles in any polygon is the sum of four right angles.

The Greeks played with whole number (integer) divisors of larger numbers, categorizing them based on their sums. The divisors of 18 are 1, 2, 3, 6, and 9. These sum to 21, which is more than 18, and the Greeks categorized 18 as *abundant* because the sum of its divisors exceeded it.

The divisors of 27 are 1, 3, and 9. These sum to 13, which is less than 27, and the Greeks categorized 27 as *deficient* because it exceeded the sum of its divisors.

The divisors of 6 are 1, 2, and 3. These sum to 6, which is of course equal to 6, and the Greeks categorized 6 as *perfect* because the sum of its divisors matched it.

There is a special term for numbers like 220 and 284, because of the relationship of the sums of their divisors. They are called *amicable* numbers. To emphasize the rarity of amicable numbers, historians of math have pointed out that another pair of amicable numbers wasn't found until 2000 years after Pythagoras, and the numbers in this pair were five-digit numbers.

EXERCISE

Find and sum the factors of 220 and 284.

PROBLEM 7.3   Categorize 24, 25, and 28 as abundant, deficient, or perfect numbers.

PROBLEM 7.4   Categorize 140, 144, and 148 as abundant, deficient, or perfect numbers.

One of Pythagoras' disciples, named **Hippasus**, is famous for discovering irrational numbers, and it is believed that he was banished or killed because he revealed this discovery. The Pythagorean cult of numbers loved numbers, played with numbers, thought about numbers, and worshipped numbers. They believed that all the possible numbers were combinations of the natural numbers, or rational numbers, numbers that can be expressed as the ratio of two natural numbers.

The natural numbers are 1, 2, 3, 4, and as high as you'd like to go. Making ratios from these numbers gives us $\frac{1}{2}, \frac{1}{3}, \frac{1}{4}, \frac{2}{3}, \frac{3}{4}, \frac{3}{2}, \frac{4}{3}$, and so on, as many as we'd like to create. These are the rational numbers.

In considering a right triangle with legs of 1 and 1, the formula $a^2 + b^2 = c^2$ tells us that the hypotenuse is $\sqrt{2}$, since $1^2 + 1^2 = 2^2$. The question arose: What rational number is $\sqrt{2}$?

It has to be close to $\frac{7}{5}$, because $\left(\frac{7}{5}\right)^2 = \frac{49}{25}$ and $\frac{50}{25} = 2$.

It has to be close to $\frac{17}{12}$, because $\left(\frac{17}{12}\right)^2 = \frac{289}{144}$ and $\frac{288}{144} = 2$.

But nobody could find the exact pair of integers to make the exact fraction that, when squared, would equal 2. And Hippasus found a new approach to the problem and proved that there wasn't an answer.

Hippasus observed that if some fraction, $\frac{x}{y}$, where x and y are whole numbers, is $\sqrt{2}$, then x and y must themselves be odd or even numbers. There are four distinct possibilities for $\frac{x}{y}$: it fits one of these models.

$$\frac{odd}{odd} \quad \frac{odd}{even} \quad \frac{even}{odd} \quad \frac{even}{even}$$

The last possibility, $\frac{even}{even}$, can be dismissed, because two even numbers have a common factor of 2, and therefore $\frac{even}{even}$ can be reduced to one of the other three possibilities. Hippasus demonstrated logically that none of the models can be $\sqrt{2}$. The easiest one to consider is $\frac{odd}{odd}$.

If you square an $\frac{odd}{odd}$ number, you get another $\frac{odd}{odd}$ number, like squaring $\frac{7}{5}$ to get $\frac{49}{25}$. The fractional representations of 2 all require an even number in the numerator: $\frac{2}{1}, \frac{4}{2}, \frac{6}{3}, \frac{8}{4}, \frac{10}{5}$, and so on. So, no $\frac{odd}{odd}$ number, when squared, can give us a numerator that is exactly twice as big as its denominator.

EXERCISE

Show that $\dfrac{odd}{even}$ and $\dfrac{even}{odd}$ cannot be models for $\sqrt{2}$.

Why did they kill or banish Hippasus? Well, if your whole system of belief is based on whole numbers, it is quite a scandal that a simple number like $\sqrt{2}$ cannot be expressed as the ratio of two whole numbers.

The next Greek genius is **Zeno**, who thought a great deal about infinity. He formulated many entertaining paradoxes involving infinite numbers of fractions, and the resolution of these paradoxes leads to the modern approach to calculating the sum of an infinite series.

Let's say you had a project to complete, and you did half of it. And then you did half of the remaining half (so you were $\frac{3}{4}$ done). And then you did half of the remaining quarter (so you were $\frac{7}{8}$ done). And then you did half of the remaining eighth (so you were $\frac{15}{16}$ done). At this rate, would you ever get the project done?

Given an infinite amount of time, yes. Here is the arithmetic. Let S be the sum of all the fractions of the project you complete.

$$S = \frac{1}{2} + \frac{1}{4} + \frac{1}{8} + \frac{1}{16} + \cdots \text{infinitely}$$

Each term in the series is half of the previous term.

Now, multiply the previous equation by 2:

$$2S = 1 + \frac{1}{2} + \frac{1}{4} + \frac{1}{8} + \cdots \text{infinitely}$$

If the original equation is subtracted from this equation, all the fractions cancel out and we are left with the wonderfully simple expression $S = 1$. This says that the sum (of all the fractions) is 1.

EXERCISE

Show that 0.999999... (repeating forever) = 1.

PROBLEM 7.5   Find the sum of the infinite series

$$\frac{3}{7} + \frac{9}{49} + \frac{27}{343} + \frac{81}{2401} + \cdots$$

where every term after the first one is $\frac{3}{7}$ of the previous term.

PROBLEM 7.6   Find the sum of the infinite series

$$\frac{4}{5} + \frac{16}{25} + \frac{64}{125} + \frac{256}{625} + \cdots$$

where every term after the first one is $\frac{4}{5}$ of the previous term.

The Greeks were fascinated by geometry problems, and one in particular was to construct a square and a circle that were equal in area. "Construct" is an ordinary word, but the Greek geometers used it in a very specialized way. They had rules for geometrical proofs and constructions, just as we have rules for our intellectual games.

For example, if you are solving a crossword puzzle, you decide whether it is okay to check a dictionary. If you and a friend are playing chess, you and your friend decide whether it is okay for a player to take back a move. Well, in the Greek game of geometry, they agreed to limit themselves to two tools: a compass and a straight-edge. A compass is V-shaped device with a sharp point (needle) on one branch of the V and a pencil point on the other branch. If you hold it upside down and anchor the needle firmly, you can twirl the compass around to draw a circle. A straight-edge is simply an unmarked ruler. You can use it to draw straight lines, but not to make measurements (it is unmarked).

The Greeks found that they could do many things with a compass and a straightedge.

If they had an angle, they could draw some arcs and cut the angle exactly in half, bisecting it.

If they had a straight line and a point outside the line, they could draw a second line parallel to the first one that passed through the point.

If they had a line segment, they could find its midpoint. And they could draw a perpendicular line that passed through the midpoint.

They could copy and angle exactly. They could extend a line indefinitely. There were a few basic and interesting things they could do, and by combining these they could do quite complicated proofs and constructions.

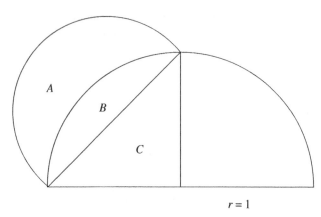

$r = 1$

FIGURE 7.9 Hippocrates made some headway on the classic problem of finding a square and a circle that had exactly the same area.

Some of the complicated things they wanted to do turned out to be impossible though, because the compass and straightedge, while versatile, weren't perfect. The problem of making a square and circle equal in area was one of those impossible challenges, but it wasn't until 2000 years after the classical Greeks struggled with it that Europeans proved that this, and a couple of similar Greek problems, simply could not be solved.

One Greek who made impressive headway on the "squaring the circle" problem was **Hippocrates**. In Figure 7.9, two areas, labeled A and C, are equal in area. Area A is completely bounded by curved lines. Area C is completely bounded by straight lines. Since a circle is completely bounded by curved lines and a square is completely bounded by straight lines, Hippocrates' diagram looks like a useful forerunner to a solution. Unfortunately, it is a special case, and it does not lead to a general method for making equal curved and straight areas.

Areas B and C together are $\frac{1}{4}$ of a circle with a radius of 1 and a diameter of 2.

Area C is a right triangle with sides of 1, 1, and $\sqrt{2}$.

Areas A and B together are half of a circle with a radius of $\frac{\sqrt{2}}{2}$.

You can use the formula for the area of a circle $\left(A = \pi r^2\right)$ to verify that the area of quarter-circle $\left(B + C\right)$ equals the area of semicircle $\left(A + B\right)$, which is $\frac{\pi}{4}$.

When we subtract area B from the quarter-circle and the semicircle, we are left with $C = A$. (Area A is called a "lune," and its shape makes people think of the moon.)

Computing areas and volumes was a huge concern to the Greeks. One of the specialists who contributed was **Democritus**, who found formulas for the volumes of pyramids and cones. It seems obvious that the volume of a cone would be smaller than the volume of the smallest cylinder you could put the cone in, but how much smaller? Was it a definite fraction of such a cylinder, or did it depend on the height and slant of the cone?

Democritus considered a cone to be a pyramid with an infinite number of very small sides on its base, making a circle. The pyramids we are most familiar with, the great pyramids of ancient Egypt, had square bases. But there is no reason a pyramid couldn't have a pentagon, hexagon, or other larger polygon as its base. The simplest possible pyramid would have a triangular base, and if the base and the three sides were identical equilateral triangles, that shape would be a special one called a *tetrahedron*. It could be rotated so that any of its sides became its base, and it would look exactly the same regardless of its orientation.

The analogy of a cone to a pyramid was useful because there are clever geometrical dissections of a rectangular solid (a box shape) that show a pyramid has a volume one third of the volume of the box it fits into.

The result was a formula you may remember from high school geometry: $V = \frac{1}{3}\pi r^2 h$, where V is the volume of the cone, r is the radius of the circle at the bottom of the cone, and h is the height of the cone.

That tetrahedron we just mentioned is an example of a very special and small group of shapes. It is a three-dimensional object that has a regular polygon as each face. A more familiar example is a *cube*: a cube has six faces, each one of which is a square.

Remarkably, there are only three more like this. An *octahedron* has eight equilateral triangles as its faces. A *dodecahedron* has 12 regular pentagons. An *icosahedron* has 20 equilateral triangles. There are some curious symmetries: the dodecahedron has 12 sides and 20 vertexes, the icosahedron has 12 vertexes and 20 sides. The cube has six sides and eight vertexes, the octahedron has eight sides and six vertexes. (The tetrahedron has four sides and four vertexes, and no partner. Some people say it's its own partner.) **Plato**, the philosopher, was enchanted by these shapes, and they are known to us as the Platonic solids. Just as Pythagoras developed a somewhat goofy cult of numbers, Plato had his somewhat goofy ideas about the Platonic solids. He thought the universe was made of them. Remember, in those days, it was popular to think there were only four elements: air, fire, earth, and water. Plato thought air was made of octahedrons, fire was made of tetrahedrons, earth was made of cubes, and water was made of icosahedrons.

Other than that, Plato was a brilliant mathematician and logician. He started a school in Athens, called the Academy, in 387 BC, where geometry was held in the highest regard, and many of the Academy's students became important mathematicians themselves.

EXERCISE

Explain why there are only five Platonic solids. (There is a good geometrical reason, having to do with the fact that there are 360° around a point.)

EXERCISE

Try to find a relationship among the number of faces, the number of vertexes, and the number of edges of a Platonic solid. (There is one.)

One of the greatest mathematicians of all time, **Euclid**, was active around 300 BC. Euclid is one of those names from antiquity, like Pythagoras and Archimedes, that everybody recognizes, and Euclid is regarded as the key figure in plane geometry.

Euclid was the author of a math book, the *Elements*, that has been a popular success for centuries. Translated into dozens of languages, it is said to have been second only to the Bible in terms of editions and copies produced. That statistic may no longer be accurate (modern mass production, and commercial successes like the Harry Potter series, may have overtaken Euclid), but the *Elements* remains relevant to this day. The book is basically a compilation of everything that was known about geometry at the time Euclid was alive, and its structure is rather incredible: Euclid sets out a small number of basic ideas (like parallel lines do not meet) and proceeds to derive from these few ideas hundreds of theorems about lines, angles, triangles, polygons, areas, transformations, circles, arcs, tangents, chords, constructions, proportions, divisors, prime numbers, number theory, and three-dimensional figures. He proves them all. *Proving* is the essential thing. If there's one thing you remember from high school geometry above all else, it is probably the process of making proofs. We deduce one fact at a time from previously acknowledged facts, building an ironclad case that what we conclude is true. Euclid was a great "prover" but among the classical Greeks some credit must be given to **Aristotle** who, though regarded as more of a philosopher than a mathematician, was one of the first and greatest geniuses to develop mathematical logic and deductive proof. (Do you know, by the way, the connection between Aristotle and Alexander the Great? It is widely said, but disputed and uncertain, that Aristotle was Alexander's tutor. It is certain, however, that Aristotle had a long friendship with King Philip of Macedon, Alexander the Great's father. Aristotle was also a student, and later a teacher, at Plato's Academy in Athens.)

Euclid contributed to many areas of math. He studied perfect numbers and found a formula to generate some. He noticed that the first four perfect numbers, 6, 28, 496, and 8128 are also triangular numbers. He showed via geometry why the algebraic formula $(a+b)^2 = a^2 + 2ab + b^2$ works. He made his own proof of the Pythagorean theorem. He developed the fundamental theorem of arithmetic that every number is the product of a unique set of prime numbers. We are going to look in more detail at two of his areas of interest: prime numbers and the golden ratio.

Prime numbers are those that cannot be factored; their only factors are themselves and 1. 13 is prime. It is 13×1. 12 is not prime: it is 2×6 or 3×4.

Prime numbers are still a topic of extensive mathematical research today, one of many connections between ancient Greek mathematics and modern mathematics. There are countless unproven conjectures about prime numbers.

One is that between any number **n** and twice that number **2n** there is at least one prime. We can see that between 5 and 10 there is a prime (7), between 8 and 16 there are primes (11 and 13), and so on, but how could we prove the general case for any **n** and **2n**?

Another idea is that there is a prime between any two consecutive square integers. Between 16 and 25, we find 17, 19, and 23. Between 25 and 36, we find 29 and 31. It seems that this conjecture is true, but again nobody has been able to prove it.

Primes that differ by 2 are called twin primes. Twin primes include 11 and 13, 17 and 19, 29 and 31, and 41 and 43. It appears that there is an infinite supply of twin primes, but this too is unproven.

Euclid proved that there were an infinite number of primes, using a clever argument: he postulated that the number of primes was limited and then showed that his assumption has to be wrong.

In simplified form, the proof says: take all the primes (2, 3, 5, 7, 11, 13, 17, 19... until the last one) and multiply them all together. The resulting large number is divisible by every prime number. Now add 1 to this large number. The new number we just created is not divisible by any of the primes—dividing it by any prime will leave a remainder of 1. So this new number is prime. It is a prime number not included in our starting list of all the primes, because it is larger than the largest of those numbers. So even when we think we have all the prime numbers, there are more.

Many people have tried to find formulas to generate prime numbers. Powers of 2, minus 1 was one attempt. $2^2 - 1 = 4 - 1 = 3$. $2^3 - 1 = 8 - 1 = 7$. It's a promising start. But $2^4 - 1 = 16 - 1 = 15$. $32 - 1 = 31$, prime again, but $64 - 1 = 63$, not prime.

How about the product of consecutive even numbers, minus 1? $(2 \times 4) - 1 = 7$. Prime.

$$(4 \times 6) - 1 = 23, \text{ prime.}$$
$$(6 \times 8) - 1 = 47, \text{ prime.}$$
$$(8 \times 10) - 1 = 79, \text{ prime.} \quad \text{It looks like we're on to something.}$$
$$(10 \times 12) - 1 = 119. \quad \text{But } 119 = 7 \times 17, \text{ so our streak ends.}$$

Maybe we can tinker with our theory and say that the product of consecutive even numbers minus 1 *usually* is a prime number?

$$(12 \times 14) - 1 = 167. \text{ Prime.}$$
$$(14 \times 16) - 1 = 223. \text{ Prime.}$$

How can we quickly determine whether 167 and 223 are prime? At first glance, testing for divisors looks tedious: don't we have to see if 167 is divisible by *every* number less than 167? No we don't. In fact, we only need to test a few numbers. If there are factors of 167, at least one of them will be less than (or equal to) the square root of 167.

Think about the numbers that multiply to 100:

$$2 \times 50$$
$$4 \times 25$$
$$5 \times 20$$
$$10 \times 10$$

Each pair (except 10×10) contains a number less than 10 and a number more than 10. 10 is the square root of 100. If we multiplied two numbers that were both less than 10, we'd get a number that was less than 100. If we multiplied two numbers that were both more than 10, we'd get a number that was more than 100. So, for any pair of unequal factors, one must be more than the square root and one must be less than the square root.

Let's go back to 167. To test whether 167 is prime, we first take its square root. 169 is 13^2, so the square root of 167 is less than 13. If 167 can be factored, one of its factors is a prime number less than 13. We only need to test 3, 5, 7, and 11, and the tests for 3 and 5 can be done in a flash. If a number is divisible by 3, its digits add up to a multiple of 3. 167 has digits that add up to 14, so it is not divisible by 3. Even if you didn't know this gimmick, it wouldn't take you long with a calculator to see that 167 is not evenly divisible by 3. We can also dismiss 5 as a factor or 167. Multiples of 5 end with 0 or 5 as their last digit. So we try 167 divided by 7, and 167 divided by 11, and we see very quickly that 167 is a prime number.

How about 223? The square root of 223 is less than 15, so the prime factors we need to test are 3, 5, 7, 11, and 13. In less than a minute, we can see that 223 is prime.

So, the general formula of *the product of consecutive even integers minus 1* seems to often, but not always, give a prime number.

As an exercise, you might try to develop a formula to generate prime numbers. You would be joining the ranks of countless ancient Greeks and modern mathematicians.

Euclid made a key calculation about a geometric ratio currently known as phi, a letter of the Greek alphabet written as φ, and also known as the *golden ratio*. If you examine any rectangle, you can make a ratio between its width and its length. The Greeks thought rectangles with one particular ratio were especially beautiful, and they called these rectangles *golden rectangles*. You can see golden rectangles in their art and architecture, and many modern products are also in the same shape.

If the ratio of width to length matches the ratio of length to (width *plus* length), the rectangle is a golden one. The value of this ratio, φ, is about 1.618, and Euclid found it by algebra. The exact value is $\dfrac{1+\sqrt{5}}{2}$.

φ shows up in many surprising places. Euclid found it in a regular pentagon. If you have a regular pentagon with sides all equal to 1, the diagonal that connects any two vertexes across the pentagon has a length of $\dfrac{1+\sqrt{5}}{2}$.

In Figure 7.10, *ABCDE* is the pentagon, with each side = 1, and *AC* is its diagonal. *F* is the midpoint of *AE*, and *G* and *H* are points on *AC* that form rectangle *GHDE*.

EXERCISE

Prove that triangles *ACF* and *AGE* are similar triangles with angles of 72°, 18°, and 90°, and use the ratios of the similar triangles to show that the length of *AC* is φ.

Around AD 1200, an Italian genius known as both Fibonacci and Leonardo of Pisa worked extensively on a series of numbers generated by a simple rule: the next number is the sum of the two previous numbers. His series starts: 1, 1, 2, 3, 5, 8, 13, 21, 34, 55, 89, 144, 233, 377. What does this have to do with φ and the Greeks?

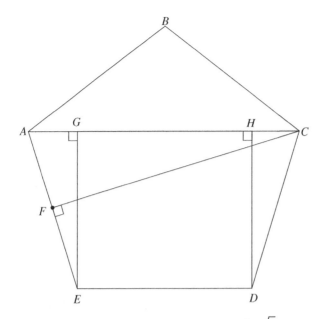

FIGURE 7.10   The diagonal of a regular pentagon is exactly $\dfrac{1+\sqrt{5}}{2}$; times longer than the side of the pentagon.

Well, if you keep calculating the ratio of consecutive terms of Fibonacci's series, the higher you go the closer the ratio gets to φ.

$$\frac{8}{5} = 1.60$$

$$\frac{34}{21} = 1.619$$

$$\frac{377}{233} = 1.618025751$$

To nine decimal places, $\varphi = 1.618033989$.

Not long after Euclid came one of the towering figures of mathematics: **Archimedes**. His genius and the breadth of his interests are overwhelming. Just as an entire book could be written about Euclid's mathematics or a full year's college course could focus on Euclid, an entire book or a full year could easily be devoted to Archimedes. Archimedes calculated π better than anyone had ever done before. He found the volume and surface area of a sphere. He calculated $\sqrt{3}$ better than anyone had ever done. He found a method of computing curved areas by treating them essentially as an infinite number of thin rectangles; he was essentially doing calculus 2000 years before Newton and Leibniz invented calculus.

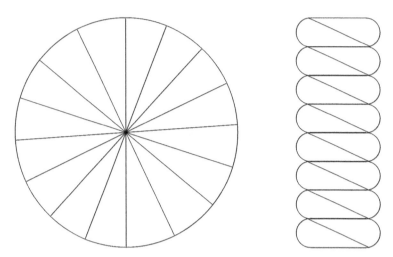

FIGURE 7.11   Archimedes carved up a circle to compute pi.

He used the same method to confirm Democritus' formula for the volume of a cone. Slicing the cone horizontally into many, many pieces made it roughly equivalent to an infinite number of ever-so-slightly-smaller circles stacked on top of each other. If each of these circles was treated like a very short cylinder (whose volume could be calculated), the stack of circles would have the same volume as the cone. Democritus himself may have thought of this 150 years before Archimedes, but since his books were lost, it is not clear whether he put all the relevant ideas together.

We'll look at one wonderful calculation Archimedes made.

On the left of Figure 7.11, a circle is cut into many slices, like a pizza. On the right side, the slices are stacked with the crust alternately on either side.

Each slice of the circle consists of two straight lines equal to the radius of the circle, and a tiny bit of the curved circumference of the circle. Archimedes imagined an infinite number of very thin slices. If he stacked them up, as shown earlier, they would form a rectangle. The width of this rectangle would be the radius of the circle, and the height of the rectangle would be half the circumference of the circle.

Since Archimedes had previously shown that the circumference of a circle is $2\pi r$, the area of the rectangular stack had to be the radius · (πr), or πr^2. Since all the slices in the rectangle came from rearranging the circle, the area of the circle also had to be πr^2, a formula we are all familiar with.

Not long after Euclid, **Eratosthenes** researched the pattern of primes, to try to find a way to predict which numbers would be prime. He observed that every second number is a multiple of 2, every third number is a multiple of 3, every fifth number is a multiple of 5, and so on. His famous "sieve" winnowed out the numbers that weren't prime, but still did not solve the puzzle. Eratosthenes did, however, do something else remarkable that ensures his fame: he measured the size of the earth.

Eratosthenes was the librarian at the great library in Alexandria, Egypt, and it occurred to him that the sun's rays could help him calculate a massive triangle. Nowadays, we tend to think that ancient people thought the earth was flat and that Columbus was alone in his confidence that the earth was round, but the Greeks had a very strong feeling that the earth was curved. Eratosthenes measured the distance between two Egyptian cities and measured the angle that the shadow of the sun's rays made at the same time on the same day in the two cities. Figuring that the sun was a point far, far away and that the two Egyptian cities were points, and assuming the earth was a sphere, Eratosthenes used trigonometry to find what part of the sphere's circumference the line connecting the two cities was. And he got it nearly right! Tragically, his brilliance was forgotten during the Dark Ages, and Europeans had to rediscover what he calculated. His map of the world, though, showing only Europe, Asia, and Africa, survives.

Another astronomer/mathematician who worked at the library in Alexandria was **Ptolemy**. Ptolemy's charts of the stars were widely studied and reproduced, and his spherical geometry is practical and profound; but he, like many of the Greek mathematicians, dabbled in all sorts of mathematical areas and also found things that had no practical value but just make us admire their cleverness. One of Ptolemy's inspirations concerned cyclic quadrilaterals.

What is a cyclic quadrilateral? A quadrilateral is a four-sided polygon. Squares and trapezoids are quadrilaterals, but a quadrilateral can be quite irregular, with no equal sides or angles. If you can inscribe a quadrilateral in a circle, so that each of its four vertexes touches the circle, it is a cyclic quadrilateral. Some quadrilaterals are cyclic, some are not. Here is Ptolemy's brilliant but useless observation: in a cyclic quadrilateral, the sum of the products of the pairs of the opposite sides is equal to the product of the diagonals. What? Look at Figure 7.12.

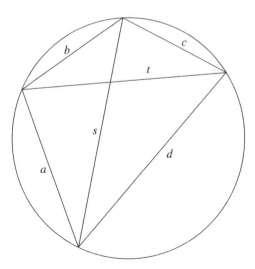

FIGURE 7.12   Ptolemy discovered a relationship between the sides and diagonals of a quadrilateral inscribed in a circle.

The sides of the quadrilateral are **a**, **b**, **c**, and **d**. The diagonals are **s** and **t**. Ptolemy figured out (and proved, of course) that $(\mathbf{a}) \cdot (\mathbf{c}) + (\mathbf{b}) \cdot (\mathbf{d}) = (\mathbf{s}) \cdot (\mathbf{t})$. It is unknown whether anyone has ever put this fact to a practical use.

Another Greek who combined useful insight and fantastic irrelevance was **Apollonius**. If you loved or hated circles, ellipses, parabolas, and hyperbolas during high school, you can thank or blame Apollonius for a fair portion of your math curriculum.

The conic sections have enormous practical value. The orbits of the planets around the sun are ellipses, and NASA uses this fact when it sends spacecraft to rendezvous with the planets. The trajectories of artillery shells are parabolas. Cameras that produce 360° views have hyperbolic mirrors. Engineers could give you dozens of other examples. Apollonius described all the essential features of the conic sections, starting with the most basic: how the intersection of a cone and plane is a conic section. Let the plane be parallel to the circle at the base of the cone, and the intersection is a circle. Tilt the plane a little bit, and the intersection is an ellipse. Tilt the plane even more, so the plane is parallel to the side of the cone, and the intersection is a parabola. Tilt the plane still more, so it is perpendicular to the base of the cone, and the intersection is a hyperbola.

Conic sections have their unique physical definitions and unique algebraic formulas, relating characteristics of points that are on them. In a circle, there is a simple center, and every point on the circumference of the circle is a certain distance (the radius) away from the center. In an ellipse, there are two points, called foci, that somewhat share the duties of being the "center." The sum of the distances to each focus is the same for every point on an ellipse. The hyperbola twists the ellipse's rule around: instead of the sum, it is the difference between the distances to the foci that is the same for every point. The parabola has a special wrinkle: every point on the parabola is the same distance away from a point (the focus) and a line (the directrix). The conic sections come in varieties. Ellipses can be nearly circular (the foci are close) or very elongated (the foci are far apart). Parabolas can be wide or narrow. There are ways to compute areas of conic sections, draw tangents to them, and construct them.

Apollonius toiled on all these things, which are useful, and then also toiled on obscure and whimsical subjects related to the useful subjects. For example, imagine that you had one circle, and you wanted to draw three more circles that just touched the first one; in other words, these three new circles were tangent to the first one. No big deal. Easy. Draw a circle. Pick three points on that circle, and draw additional circles that pass through those three points. But turn the problem around and it becomes a real challenge: start with three circles drawn somewhere in a plane, and now try to find one circle that is tangent to the other three (see Figure 7.13).

Most people wouldn't even be confident about where to begin on this problem. Apollonius showed how it could be done, and in a virtuoso performance showed all sorts of variations, depending on whether you preferred your new circle inside or

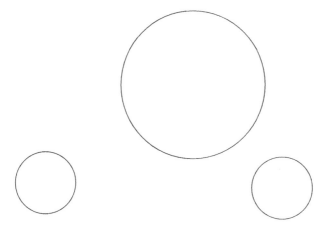

FIGURE 7.13   Apollonius analyzed the problem of finding a circle that was tangent to three other circles.

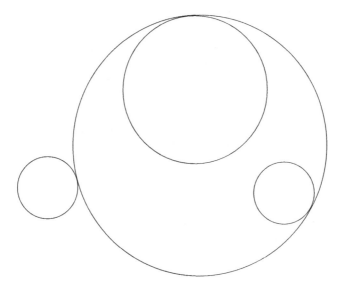

FIGURE 7.14   Apollonius solved the problem of finding a circle that was tangent to three other circles.

outside the starting three, or a mix (like two inside and one outside). It is doubtful whether anyone has ever put his technique to a practical use.

Apollonius also worked very hard on one of the great (impossible) problems of ancient Greece: constructing a cube that had exactly twice the volume of another cube.

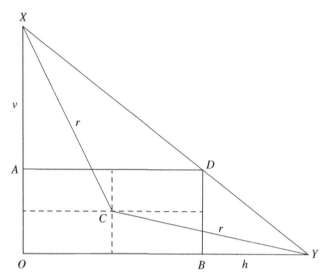

FIGURE 7.15 Apollonius came close to solving the classic problem of duplicating the cube.

Let's say you had a cube with a side of 2. Its volume would be $2 \times 2 \times 2$, or 8. A cube with exactly double the volume would have a volume of 16, and its side would be the cube root of 16 (just as 2 is the cube root of 8). The cube root of 16 is two times the cube root of 2. So the problem of "duplicating the cube" essentially comes down to constructing the cube root of 2.

Apollonius drew a rectangle, as in Figure 7.15, found its center, extended its sides, and drew a circle that had the same center as the rectangle and intersected the extensions of the rectangle in such a way that the two points of intersection and another corner of the rectangle were all on the same line. Whew! Wonderfully, two line segments in this complicated drawing were in the ratio of 1 to the cube root of 2.

The only problem was a technical one involving the famous compass and straight-edge. While Apollonius was algebraically correct, that the special circle he described could not be constructed precisely with just the compass and straightedge.

EXERCISE

If rectangle *OADB* in Figure 7.15 has a length of 4 and a height of 2, and *v* is the vertical extension of *OA* to *X*, and *h* is the horizontal extension of *OB* to *Y*, and the distance from the center of the rectangle *C* to both *X* and *Y* is *r*, and *X*, *D*, and *Y* are collinear, then show that the length of *h* is $\sqrt[3]{16}$.

Another amazing Greek mathematician who was interested in cube roots was **Heron**. Heron came up with a formula to compute them:

$$a + \frac{b\,d}{(b\,d + a\,D)} \cdot (b - a)$$

where $a^3 < N < b^3$, and $d = N - a^3$, and $D = b^3 - N$

That is, **N** lies somewhere between two known perfect cubes: **a** and **b**. **D** and **d** are how much more or less than **N** these cubes are.

Let's see how this works with an example. Let's compute the cube root of 100. We know it has to be between 4 and 5, because 4^3 is 64, and 5^3 is 125.

In Heron's formula, **a** is 4 and **b** is 5, because $64 < 100 < 125$ ($\mathbf{a^3 < N < b^3}$)

This makes **d** $= 100 - 64 = 36$.

And it makes **D** $= 125 - 100 = 25$.

We can substitute **a, b, d,** and **D** into the formula:

$$4 + \frac{(5)(36)}{\left[(5)(36) + (4)(25)\right]} \cdot (5 - 4)$$

The result is 4.642857, which is awfully close.

You might wonder why the term $(\mathbf{b} - \mathbf{a})$ is necessary at the end of Heron's cube root formula—wouldn't $(\mathbf{b} - \mathbf{a})$ always be 1? Yes, if we select *integer* cubes above and below **N**. But it is not necessary to limit ourselves to integers. Someone tinkering with numbers might have observed that $(4.5)^3$ is $91\frac{1}{8}$ and $(4.8)^3$ is $110\frac{74}{125}$. That would indicate that $\sqrt[3]{100}$ is between 4.5 and 4.8. The appropriate substitutions in Heron's cube root formula would be

$$4.5 + \frac{(4.8)\left(100 - 91\frac{1}{8}\right)}{\left[(4.8)\left(100 - 91\frac{1}{8}\right) + (4.5)\left(110\frac{74}{125} - 100\right)\right]} \cdot (4.8 - 4.5)$$

This reduces to 4.6415847 that is even closer to the actual cube root of 100, 4.6415888

PROBLEM 7.7   Use the formula to approximate the cube root of 444.

PROBLEM 7.8   Use the formula to approximate the cube root of 44,400.

Heron appears to be the first person to have thoroughly documented a foolproof way to compute square roots as precisely as you want to. His method starts with a guess, and the guess does not have to be particularly good. The "guess" divided into the starting number gives a "result," and then the guess and the result are averaged. This average becomes the next guess, and the procedure is repeated until the guess and the result are the same.

Let's say we wanted the square root of 700, and we guessed that it was 20.

$$\frac{700}{20} = 35. \text{ So we average 20 and 35 and guess again. } \frac{20 + 35}{2} = 27.5$$

$$\frac{700}{27.5} = 25.454545 \qquad \frac{27.5 + 25.454545}{2} = 26.477273$$

$$\frac{700}{26.477273} = 26.437768 \qquad \frac{26.477273 + 26.437768}{2} = 26.457521$$

$$\frac{700}{26.457521} = 26.457505 \qquad \frac{26.457521 + 26.457505}{2} = 26.457513$$

We can go on as far as we'd like, but already $(26.457513)^2 = 699.9999941$

PROBLEM 7.9   Use this method to calculate the square root of 73, correct to four decimal places. Your first guess can be in the ballpark or it can be ridiculous—the formula is self-correcting.

PROBLEM 7.10   Use this method to calculate the square root 288, starting with the guess of 15.

Heron undoubtedly was aware of the Babylonian techniques for square roots and cube roots. It is widely believed that his methods were their methods, and that his contribution was to document and explain and prove what had previously only been asserted.

His iterative process for square roots also works for cube roots, although it works a little more slowly. To find the cube root of 100, we might "guess" 4, and divide 100 by 4 twice.

$$\frac{100}{4} = 25 \qquad \frac{25}{4} = 6.25$$

Now we average 4, 4, and 6.25 to get 4.75. Repeating the process…

$$\frac{100}{4.75} = 21.0526 \qquad \frac{21.0526}{4.75} = 4.4321$$

Now we average 4.75, 4.75, and 4.4321 to get 4.6440

In this way we get closer and closer to the true value.

Heron was a busy man. He invented machines, toys, and mechanical devices. He worked on physics problems and calculated the pressure in moving fluids. The thing he is most famous for, however, is a remarkable formula for the area of a triangle.

Now, you might think this is an irrelevance (and for some triangles it is). After all, there was a perfectly good and very easy formula already known for computing the area of a triangle: Area $= \frac{1}{2}$ base $\cdot$ height. There was even a second formula known, from trigonometry, which could be used when you knew the lengths of two sides and the angle in between them: Area $= \frac{1}{2} a \cdot b \cdot \sin C$. However, for some triangles you might not know the height (needed for the first formula) or the angle (needed for the second formula), but if you knew the lengths of all three sides, then you could use Heron's formula.

$$\text{Area} = \sqrt{(s) \cdot (s-a) \cdot (s-b) \cdot (s-c)}$$

Heron's formula says that the area is equal to the square root of four lengths multiplied together. The four lengths are as follows: the semi-perimeter of the triangle, or $\frac{a+b+c}{2}$, called s, and the values $(s-a), (s-b)$, and $(s-c)$, where a, b, and c are the sides of the triangle.

How this idea came to Heron is unknown, because there is no obvious relationship between the perimeter of a triangle (or half the perimeter) and its area, but Heron's work is very well documented and preserved, and what we know is that he developed this formula after proving all sorts of other facts about triangles and circles and polygons that look quite unrelated to the task. In a modern textbook, about six pages are required to demonstrate and prove all of Heron's apparently unrelated steps. At the end, through some clever substitutions and analogies, Heron derives his formula.

Let's try it with a simple triangle, where we know the area easily from the standard formula. In the right triangle with sides of 5, 12, and 13 (a famous Pythagorean triple), the sides of 5 and 12 meet at a right angle, so we can say the base is 12 and the height is 5. This makes the area 30.

The first requirement of Heron's formula is the semi-perimeter. $\frac{5+12+13}{2} = 15$.
Next we calculate $(s-a), (s-b)$, and $(s-c)$. $15-5=10$. $15-12=3$. $15-13=2$.
Next we multiply: $15 \cdot 10 \cdot 3 \cdot 2 = 900$.
Finally, we take the square root: $\sqrt{900} = 30$.

PROBLEM 7.11 **Use Heron's formula to compute the area of a triangle with sides 13, 20, and 21. Show the work.**

PROBLEM 7.12 **Use Heron's formula to compute the area of a triangle with sides 17, 25, and 26.**

The last Greek mathematician we will look at is **Diophantus**, who is one of the giants of the ancient world. He had a strong preference for dealing with whole numbers, and he was especially fascinated by perfect squares and cubes. A branch of sophisticated arithmetic is named for him: solving Diophantine equations, where the object is to find integer solutions.

A typical equation with one unknown, $2x = 5$, has only one solution: $x = 2\frac{1}{2}$. If you like integers, you're stuck. $2\frac{1}{2}$ is the only solution.

But an equation with two unknowns has an unlimited number of solutions. $2x + y = 5$ could be solved by any of the following pairs:

$$x = 0 \qquad y = 5$$
$$x = \frac{1}{2} \qquad y = 4$$
$$x = 1 \qquad y = 3$$
$$x = 1\frac{1}{2} \qquad y = 2$$
$$x = 2 \qquad y = 1$$
$$x = 2\frac{1}{2} \qquad y = 0$$
$$x = 3 \qquad y = -1$$

and so on. If you have a preference for whole numbers, and especially positive integers, the solutions that interest you are ($x = 1$ and $y = 3$), and ($x = 2$ and $y = 1$). The other solutions contain zero or a fraction or a negative number.

Determining whether an equation had integer solutions was a foundation of Diophantus' work. An equation might have no integer solutions, one, several, or an infinite number of them. We have already seen some Diophantine equations in this book. The Chinese problem where the man buys 100 birds for 100 coins is one. Nobody is selling fractions of hens in this problem—we are dealing only with whole numbers.

The possibility for multiple solutions rests on the relationship between the numbers in the problem. $3x + 5y = 50$ features the numbers 3 and 5, which multiply to 15, which means 15 can be thought of as either $3+3+3+3+3$ or $5+5+5$, and one sum can replace the other. A simple solution to $3x + 5y = 50$ is: $x=0$ and $y = 10$. (The check says $0 + 50 = 50$.)

But we can replace 3 y's with 5 x's, giving the solution: $x = 5$ and $y = 7$ ($15 + 35 = 50$). We can replace again: $x = 10$ and $y = 4$ ($30 + 20 = 50$). And again: $x = 15$ and $y = 1$ ($45 + 5 = 50$).

So, there are 4 whole number solutions to $3x + 5y = 50$, the pairs (0, 10), (5, 7), (10, 4), and (15, 1).

There are an infinite number of other solutions (as long as we're willing to use fractions and negative numbers).

If you don't mind negative numbers, (20, −2) is a solution: ($60 + [-10] = 50$).

If you don't mind fractions, $\left(6\frac{2}{3}, 6\right)$ is a solution: ($20 + 30 = 50$).

Diophantus himself, even though his name is now associated with equations with strictly whole number solutions, presented a large number of problems in his book *Arithmetica* that he solved using rational fractions. Sometimes, despite our hopes, there just isn't a whole number solution.

It is not always easy to tell if an equation will have a Diophantine (whole number) solution. Here is an equation which obviously does not: $5x + 10y = 36$. Do you see why not? Whole numbers multiplied by 5 will end in the digit 5 or the digit 0; whole numbers multiplied by 10 can only end with the digit 0. The sum of $5x$ plus $10y$ here can only end in a 5 or a 0, and the equation requires a final digit of 6 because the sum equals 36.

Let's say we were looking for Diophantine solutions to $2x + 13y = 56$. We could proceed by trial and error until we found a first solution, but let's take a very simple idea—what is the largest multiple of 13 that is less 56? It is 52. ($4 \times 13 = 52$). The difference between 56 and 52 is fortuitously a multiple of the other coefficient in this equation, 2, so we can see that the (x, y) pair $(2, 4)$ is a solution to $2x + 13y = 56$ ($4 + 52 = 56$).

What happens if we take away one of the four 13s? We cannot replace a 13 by a whole number of 2s, so there will not be a Diophantine solution like $(x, 3)$ where x is a whole number. But we can replace two 13s by thirteen 2s, so there is a solution like $(x, 2)$ where x is a whole number. That solution is $(15, 2)$, which we can verify because $30 + 26 = 56$. A subsequent replacement gives us a third solution: $(28, 0)$.

What if we modify the starting equation ever so slightly: $2x + 13y = 57$.

A slightly different approach is to subtract 13 from 57 until we get a number that is divisible by 2. It happens right away. $57 - 13 = 44$. So, a Diophantine solution to $2x + 13y = 57$ is $(22, 1)$. The check is $44 + 13 = 57$.

And we can replace blocks of 26 (2×13) as we did previously. Another solution is $(9, 3)$ because $18 + 39 = 57$. Compared to the first solution, $(22, 1)$, the number of 2s went down by 13 (from 22 to 9), and the number of 13s went up by 2 (from 1 to 3).

$(22, 1)$ and $(9, 3)$ are the only Diophantine solutions here, because we cannot subtract another 13 from 9 without getting a negative result.

PROBLEM 7.13   Find the whole number solutions of $4x + 7y = 99$.

PROBLEM 7.14   Find the whole number solutions of $5x + 12y = 299$.

A Diophantine riddle goes like this: four numbers, taken three at a time, give these four sums: 78, 94, 96, and 119. What are the four numbers?

This looks like something that could kill a whole afternoon if all you could do is guess and check. But there is a nice algebraic strategy. Each of the four sums (78, 94, 96, and 119) is missing exactly one of the numbers, so the sum of the four sums is missing each number once, and the sum of the four sums is therefore three times the sum of the four numbers. That may sound confusing. Let's simplify.

Call the four numbers A, B, C, D, with A the smallest and D the largest.

The riddle says essentially:

$$
\begin{array}{ccccccccc}
A & + & B & + & C & & & = & 78 \\
A & + & B & & & + & D & = & 94 \\
A & & & + & C & + & D & = & 96 \\
& & B & + & C & + & D & = & 119 \\
\end{array}
$$

The first sum (78) is missing the largest original number (D). The fourth sum (119) is missing the smallest original number (A).

We can add the columns of equations to discover something useful:

$$
\begin{array}{ccccccccc}
A & + & B & + & C & & & = & 78 \\
A & + & B & & & + & D & = & 94 \\
A & & & + & C & + & D & = & 96 \\
& & B & + & C & + & D & = & 119 \\
\hline
3A & + & 3B & + & 3C & + & 3D & = & 387
\end{array}
$$

We can divide 387 by 3 to find the sum of the original four numbers: 129.
The four unknown numbers must be 129 minus each of the four sums.

$$129 - 119 = 10$$
$$129 - 96 = 33$$
$$129 - 94 = 35$$
$$129 - 78 = 51$$

You can check that these numbers, taken three at a time, give the four sums in the riddle.

Notice that the differences among the original numbers are echoed in the differences in the sums:

the gaps between 10, 33, 35, and 51 are: 23, 2, and 16.

the gaps between 78, 94, 96, and 119 are: 16, 2, and 23.

PROBLEM 7.15   Four numbers, taken three at a time, give these four sums: 49, 65, 71, and 79. What are the four numbers?

PROBLEM 7.16   Four numbers, taken three at a time, give these four sums: 209, 365, 432, and 449. What are the four numbers?

Arithmetica features some very intricate problems that involve manipulating numbers in multiple ways. In one problem, 100 is to be divided three times into two whole numbers, so that the larger number in one pair compared to the smaller number in a different pair is a specific ratio. The three divisions are as follows: $(84 + 16)$, $(72 + 28)$, and $(64 + 36)$. Observe these remarkable ratios:

$$84 : 28 = 3 : 1$$
$$72 : 36 = 2 : 1$$
$$64 : 16 = 4 : 1$$

In a different challenge, two whole numbers are to donate parts of themselves to the other number to achieve certain ratios. If we call the numbers A and B, the ratio of $A + 30$ to $B - 30$ is supposed to be $2 : 1$, and the ratio of $B + 50$ to $A - 50$ is supposed to be $3 : 1$. Diophantus found 98 and 94 for A and B.

$$(98 + 30) : (94 - 30) = 2 : 1$$
$$(94 + 50) : (98 - 50) = 3 : 1$$

Only a few pages later, Diophantus is looking for four numbers which, after the first gives one third of itself to the second, and the second gives one fourth of itself to the third, and the third gives one fifth of itself to the fourth, and the fourth gives one sixth of itself to the first, with all the generous donations occurring simultaneously, the result is that the four quantities are suddenly equal. He finds: 150, 92, 120, and 114.

150 gives away 50 but receives 19, becoming 119.

92 gives away 23 but receives 50, becoming 119.

120 gives away 24 but receives 23, becoming 119.

And 114 gives away 19 but receives 24, becoming 119.

Here are some of the results Diophantus found when he investigated squares and cubes.

RESULT 1:
41, 80, and 320 when added together equal a perfect square, and any two of them together are also a perfect square. Specifically,

$$41 + 80 + 320 = 441 = 21^2$$
$$41 + 80 \qquad = 121 = 11^2$$
$$41 + 320 \qquad = 361 = 19^2$$
$$80 + 320 \qquad = 400 = 20^2$$

RESULT 2: $\qquad 40^2 + 96^2 = 104^2 \text{ (a Pythagorean triple) and}$
$$(104 - 40) = 64 = 4^3$$
$$(104 - 96) = 8 = 2^3$$

RESULT 3: $\qquad 481 = 20^2 + 9^2 \ (400 + 81), \text{ and also}$
$$481 = 16^2 + 15^2 \ (256 + 225)$$

This is a specific case of a general formula:
$$\left(a^2 + b^2\right)\left(c^2 + d^2\right) = \left(ac + bd\right)^2 + \left(ad - bc\right)^2$$
$$\text{where } a = 3, b = 2, c = 1, d = 6$$

RESULT 4: $\qquad 1105 = 33^2 + 4^2$
$$1105 = 32^2 + 9^2$$
$$1105 = 31^2 + 12^2$$
$$1105 = 24^2 + 23^2$$

RESULT 5:

The prime factors of 1105 are 5, 13, and 17.

$$5 \times 13 = 65 = \left(8^2 + 1^2\right) \text{ or } \left(6^2 + 5^2\right)$$
$$5 \times 17 = 85 = \left(9^2 + 2^2\right) \text{ or } \left(7^2 + 6^2\right)$$
$$13 \times 17 = 221 = \left(14^2 + 5^2\right) \text{ or } \left(11^2 + 10^2\right)$$

Diophantus found this curious way to construct a perfect square, given two random values. We're going to pick a random value called n, square it, and pick a random value less than n^2, which we will call a. Three other values (m, x, and y) will depend on n and a. Ultimately, we will say that if $m = (n+1)$, and $(x+a) = m^2$, and $(y+a) = n^2$, then $(xy+a)$ is a perfect square.

Let $n=4$ (a random pick). n^2 is now 16. Let $a=7$ (a random pick less than 16). Now $m=5$, and $m^2=25$. All this forces $x=18$ and $y=9$.

$$\left(xy+a\right) = \left(162 + 7\right) = 169 = 13^2$$

A second example:

Let $n=10$, and $a=44$. This forces $m=11$, $x=77$, and $y=56$.

$$\left(xy+a\right) = \left(4312 + 44\right) = 4356 = 66^2$$

Diophantus also found this recipe for making squares. If $x = m^2$, $y = (m+1)^2$, and $z = 2(x+y+1)$, then the following six numbers are all perfect squares:

$$xy + x + y$$
$$yz + y + z$$
$$zx + z + x$$
$$xy + z$$
$$yz + x$$
$$zx + y$$

An example: let $m = 2$. This forces $x = 4$, $y = 9$, $z = 28$.

$$xy + x + y = \left(4\right)\left(9\right) + 4 + 9 = 36 + 4 + 9 = 49 = 7^2$$
$$yz + y + z = \left(9\right)\left(28\right) + 9 + 28 = 252 + 9 + 28 = 289 = 17^2$$
$$zx + z + x = \left(28\right)\left(4\right) + 28 + 4 = 112 + 28 + 4 = 144 = 12^2$$
$$xy + z = \left(4\right)\left(9\right) + 28 = 36 + 28 = 64 = 8^2$$
$$yz + x = \left(9\right)\left(28\right) + 4 = 252 + 4 = 256 = 16^2$$
$$zx + y = \left(28\right)\left(4\right) + 9 = 112 + 9 = 121 = 11^2$$

Another challenge was to find three numbers such that when you added any two of them (or all of them) to 3, the result would be a square. Diophantus found 33, 64, and 189.

$$3 + 33 + 64 = 100 = 10^2$$
$$3 + 33 + 189 = 225 = 15^2$$
$$3 + 64 + 189 = 256 = 16^2$$
$$3 + 33 + 64 + 189 = 289 = 17^2$$

Diophantus could go on like this all day.

He found two fractions (which were squares) that yielded other squares when either of them was added to their product:

$$\left[\left(\frac{3}{4}\right)^2 \cdot \left(\frac{7}{24}\right)^2\right] + \left(\frac{3}{4}\right)^2 = \left(\frac{25}{32}\right)^2$$

$$\left[\left(\frac{3}{4}\right)^2 \cdot \left(\frac{7}{24}\right)^2\right] + \left(\frac{7}{24}\right)^2 = \left(\frac{35}{96}\right)^2$$

He found three mixed numbers in arithmetical progression that gave squares when any two of them were added:

$$120\frac{1}{2} + 840\frac{1}{2} = 961 = 31^2$$

$$120\frac{1}{2} + 1560\frac{1}{2} = 1681 = 41^2$$

$$840\frac{1}{2} + 1560\frac{1}{2} = 2401 = 49^2$$

And coincidentally $\left(840\frac{1}{2} - 120\frac{1}{2}\right) = \left(1560\frac{1}{2} - 840\frac{1}{2}\right)$

He found two fractions whose sum was the same as the sum of their cubes:

$$\left(\frac{5}{7}\right) + \left(\frac{8}{7}\right) = \left(\frac{5}{7}\right)^3 + \left(\frac{8}{7}\right)^3$$

Diophantus suggested that for any sum of rational cubes, there is a corresponding difference of rational cubes. In other words, for $a^3 + b^3$, there exists $c^3 - d^3$.

Taking the charming small sum of not only rational but integer cubes,

$$1^3 + 6^3 + 8^3 = 9^3 \qquad \left(1 + 216 + 512 = 729\right)$$

he found these relationships between sums and differences by algebraic manipulation:

$$1^3 + 6^3 = 9^3 - 8^3$$
$$1^3 + 8^3 = 9^3 - 6^3$$
$$6^3 + 8^3 = 9^3 - 1^3$$

Based on his findings for particular problems and challenges, Diophantus proposed many general rules, but he did not prove them. Or his proofs are lost–we know he wrote three books, but one is completely lost and less than half of the others survive. A lot of what Diophantus suggested was proved by French and Swiss geniuses in the 17th and 18th centuries.

The extant part of his writings, and the writings of later Greek and Arab mathematicians who were commenting on his work, solidify his reputation as a brilliant theorist and problem-solver. Just as the work of Euclid and Archimedes is the basis for geometry as we know it, Diophantus' work provides the basis for two branches of mathematics: algebra and number theory.

EXERCISE

Find two cubes that add up to 12^3–10^3. (Give yourself plenty of time.)

Suggestions for book or Internet research		
History, archeology	Ancient Greek democracy Greek city-states Athens, Sparta, Delphi, Corinth, Thebes The Peloponnesian War Pericles Alexander the Great	
Religion, culture	Greek mythology The *Iliad*, the *Odyssey* Socrates Hellenistic Art	
Mathematicians	Apollonius Archimedes Aristotle Democritus Diophantus Eratosthenes Euclid Eudoxus Heron Hipparchus	Hippasus Hippocrates Hypatia Pappus Plato Ptolemy Pythagoras Thales Theon Zeno
Mathematical topics	Attic Greek numbers Geometry Prime numbers Platonic solids The Academy Doubling the cube Diophantine equations	Ionic Greek numbers Conic sections Perfect numbers Semi-Platonic solids The *Elements* Squaring the circle The golden ratio

ANSWERS TO PROBLEMS

7.1

It would be tedious to draw 400 dots in a 20×20 array, but we know the 20 dots along the biggest diagonal split the remainder of the dots equally. So, one of the triangular numbers is the sum of all the numbers from 1 to 19, and the other is the sum of all the numbers from 1 to 20. $400 - 20 = 380. \dfrac{380}{2} = 190.$

So the two triangular numbers are 190 and 210.

7.3

Divisors of 24: 1, 2, 3, 4, 6, 8, 12. Sum = 36. Abundant.
Divisors of 25: 1, 5. Sum = 6. Deficient.
Divisors of 28: 1, 2, 4, 7, 14. Sum = 28. Perfect.

7.5

$$S = \frac{3}{7} + \frac{9}{49} + \frac{27}{343} + \frac{81}{2401} + \cdots$$

We need to create a situation where the fractions cancel out.

Multiplying both sides of the equation by $\dfrac{7}{3}$ gives

$$\frac{7}{3}\, S = 1 + \frac{3}{7} + \frac{9}{49} + \frac{27}{343} + \frac{81}{2401} + \cdots$$

Subtracting the original equation from this one gives

$$\frac{4}{3}\, S = 1$$

Therefore, $S = \dfrac{3}{4}$

Take your calculator and add $\dfrac{3}{7} + \dfrac{9}{49} + \dfrac{27}{343} + \dfrac{81}{2401}$. You will see that you are getting very close to $\dfrac{3}{4}$ already. There are an infinite number of terms to add still, but each of them is very small, and they get progressively smaller.

7.7

In the formula, **a** is 7 and **b** is 8, because $343 < 444 < 512$.
This makes **d** $= 444 - 343 = 101$.

And it makes $\mathbf{D} = 512 - 444 = 68$.

We can substitute a, b, d, and D into the formula:

$$7 + \frac{8 \cdot 101}{[(8 \cdot 101) + (7 \cdot 68)]} \cdot (8 - 7)$$

The result is 7.62928 and the actual value is 7.62888

7.9

$$\sqrt{73} = 8.5440$$

7.11

The semi-perimeter $= \dfrac{13 + 20 + 21}{2}$, which is 27.

Next we calculate $(s-a)$, $(s-b)$, and $(s-c)$. $27 - 13 = 14$. $27 - 20 = 7$. $27 - 21 = 6$.
Next we multiply: $27 \cdot 14 \cdot 7 \cdot 6 = 15{,}876$.
Finally, we take the square root: $\sqrt{15{,}876} = 126$.

7.13

$4x + 7y = 99$. Observe that $4 \times 7 = 28$, so it is groups of 28 that can be interchangeably seven 4s or four 7s.

Subtract 7 from 99. The result is 92. Is this a multiple of 4? Yes. So, one solution gives us $92 + 7 = 99$. That solution is $(23, 1)$. Replacing 28s gives us additionally:

$$\begin{aligned}
(16, 5) &\qquad \text{where } 64 + 35 = 99 \\
(9, 9) &\qquad \text{where } 36 + 63 = 99 \\
(2, 13) &\qquad \text{where } 8 + 91 = 99
\end{aligned}$$

A different approach (that would of course have given us the same four answers) is to find the largest multiple of 7 that is less than 99. That's 91, which is 7×13. So the first answer we would have seen is $(2, 13)$, and we would have found the other answers by replacing 7s with 4s.

7.15

A +	B +	C		=	49
A +	B		+ D	=	65
A		+ C +	D	=	71
	B +	C +	D	=	79
$3A$ +	$3B$ +	$3C$ +	$3D$	=	264

We can divide 264 by 3 to find the sum of the original four numbers: 88.
The four unknown numbers must be 88 minus each of the four sums.

$$88 - 79 = 9$$
$$88 - 71 = 17$$
$$88 - 65 = 23$$
$$88 - 49 = 39$$

8

EARLY HINDU MATHEMATICS

Our biggest debt to the mathematicians of ancient India is their notation—the symbols we use to write numbers today (0, 1, 2, 3, 4, 5, 6, 7, 8, 9) derive directly from the Hindu symbols. They were introduced to Europe by the Arabs, and therefore they are often called the **Hindu–Arabic numbers**, as well as simply the **Arabic numbers**. We take for granted that they are most sensible way to represent numbers, but of course this is a prejudice we have because of our familiarity with these numbers. The Romans would have thought **XIV** was a more sensible way to represent **14**, and the Egyptians would have preferred their hieroglyphics, the Babylonians their wedges, the Chinese their calligraphy, and so on. To a Roman, a number like **14** might suggest 3, because it is a 1 before a 4, but it would seem peculiar since there was a perfectly logical **III** available already to represent 3.

The shapes of the Hindu numerals were easy enough for the Arabs to copy and adopt, and then also for the Europeans to copy and adopt. The most important innovation was the zero, which removed the ambiguity of some earlier positional systems and made it very easy to express the result of an operation like 6 minus 6. We can say, based on the near universal use of Hindu–Arabic numbers now, that they definitely caught on.

In considering early Indian mathematics, several main points stand out. First, we don't have the greatest archeological record for these people. They did not write on baked clay tablets or easily preserved papyrus. The buildings and streets of the city at Mohenjo Daro (by the Indus river in what is now Pakistan) are preserved, but we haven't deciphered their Harappan language, so our understanding of the early

An Introduction to the Early Development of Mathematics, First Edition. Michael K. J. Goodman.
© 2016 John Wiley & Sons, Inc. Published 2016 by John Wiley & Sons, Inc.

inhabitants of the Indus valley is limited. Texts written in Sanskrit, which mix religion, poetry, philosophy, and history, are our clues to what these people knew about and achieved in math.

Second, we can make the general statement that mathematics appears to have been for them, early on, an aid to their religion and astronomy. Deep calculations are made to fix the age of cycles of the universe, and these cycles involve very large numbers of years. On an earthly level, a chief mathematical task was to design temples in a reliably repetitive and consistent way, so that a worshipper would feel equally at home in any temple. The measurements therefore had to be duplicated exactly. Perhaps there was a feeling also that there was a natural or even a perfect way to build a holy place, and the blueprints had to be precise. One indication of this is the ancient Hindu approximation for the square root of 2, which perhaps arose from the challenge of constructing a square altar exactly twice as large as another square altar.

A third thing we notice is the rediscovery or refinement of ideas that the ancient Greeks worked with, and this extends further to an overlap of interests with Babylonian and Chinese mathematics. To some extent, ancient India was isolated. Deserts and mountains separated India from Mesopotamia and Persia to the west, and the highest mountains in the world, the Himalayas, separated India from China. Still, there must have been numerous exchanges, stimulated by the activities of adventurous traders. The Hindu method of constructing a square looks very much like Euclid's method of constructing an equilateral triangle. Geometrical transformations that interested the Hindus (rectangle to square, square to circle) remind us of the classical challenges of the Greeks (squaring the circle, doubling the cube). Hindu work on angles and chords is like earlier Greek work. The Hindu formulas for square roots and cube roots are improvements on the more ancient Babylonian formulas. The content of Hindu mathematical problems and riddles, and their presentation in the form of stories, reminds us of equivalent Chinese problems.

And that is the fourth thing, and for us a very entertaining aspect of ancient Indian math: the presentation of (largely algebraic) computations in the form of fanciful stories. We will look in detail at a number of these.

THE FRUIT PROBLEM

This is a simple **systems of equations** problem, with two equations with two unknowns. The problem can be presented directly in terms of **x** and **y**, but the story form is more memorable and perhaps also a stimulus to applying the solution technique to practical everyday problems. Here is the challenge:

If you buy 7 apples and 9 citrons, it will cost you 107 cents. If you buy 9 apples and 7 citrons, it will cost you 101 cents.
   What is the individual price of 1 apple? What is the price of 1 citron?

Of course, in the original form, some other unit is assigned, not cents. The price is 107 rupees or parts of a rupee, but for our purposes we don't need to specify the units faithfully or even specify them at all.

The first thing you might notice is that in either scenario you are buying 16 pieces of fruit. By substituting 2 apples for 2 citrons, you reduce the total price by 6 cents, from 107 to 101, so it appears that an apple costs 3 cents less than a citron. This insight alone is enough to solve the problem by a series of guesses, if you have the patience or the luck to solve the problem that way. You might pick any arbitrary price for a citron (say, 10 cents) and assign a price 3 cents cheaper to an apple (here 7 cents), and compute the cost of the combination of 16 pieces of fruit in either scenario. The guess of 10 cents and 7 cents will give an answer that is too high, so you will lower your guess and try again.

The algebraic way to solve the riddle is to set up these equations, where A and C represent the prices of one apple and one citron.

$$7A + 9C = 107$$
$$9A + 7C = 101$$

The goal with systems of equations is to eliminate an unknown, so the first equation is multiplied by 9 and the second is multiplied by 7, giving this pair of equations.

$$63A + 81C = 963$$
$$63A + 49C = 707$$

Now the second equation can be subtracted from the first, with the $63A$ terms dropping out.

$$32C = 256$$

Dividing by 32 gives the price of one citron.

$$C = 8$$

You can check that replacing the symbol C by the number 8 in both of the original equations gives us the conclusion, $A = 5$. The equations are representations of the little story in the box, and you can also confirm that 8 and 5 are the prices if you work directly with the story.

If this sort of problem seems very easy to you, you can move on to the Tax Problem. If you want to test your skill with similar problems, here are several.

PROBLEM 8.1   If you buy 8 pears and 5 bananas, it will cost you 118 cents. If you buy 5 pears and 8 bananas, it will cost you 103 cents.

What is the individual price of 1 pear? What is the price of 1 banana?

PROBLEM 8.2 **If you buy 4 sweaters and 10 shirts, it will cost you $418. If you buy 7 sweaters and 8 shirts, it will cost you $570.**
What is the individual price of 1 sweater? What is the price of 1 shirt?

THE TAX PROBLEM

This type of problem requires the solver to make successive computations with fractions, but the quantity we need to operate on changes, which introduces a trap for the unwary solver.

A merchant has to give away some of his goods as a tax, as he passes through three territories. At the first stop, he has to part with $\frac{1}{3}$ of his goods; at the second stop, $\frac{1}{4}$ of what remains, and at the third stop, $\frac{1}{5}$. His total tax is 24.
What is his original quantity?

We might mistakenly add the three fractions and get $\frac{47}{60}$ and then wonder how this awkward fraction relates to 24. Does $\frac{47}{60}x = 24$? But this is misreading. Or misunderstanding, the circumstances of the problem.

The merchant does indeed pay $\frac{1}{3}$ of his goods at the first stop, but at the second stop he is paying $\frac{1}{4}$ of what he has left after he pays tax at the first stop; so he is paying $\frac{1}{4}$ of the $\frac{2}{3}$ he has left of his original inventory. Then at the third stop, he pays $\frac{1}{5}$ of what is left after the first two episodes of obnoxious taxation. He pays successively:

$$\frac{1}{3}$$

$$\frac{1}{4} \text{ of } \left(1 - \frac{1}{3}\right)$$

$$\frac{1}{5} \text{ of } \left(1 - \frac{1}{3} - \left[\frac{1}{4} \text{ of}\left(1 - \frac{1}{3}\right)\right]\right)$$

This convoluted mess becomes

$$\frac{1}{3}$$

$$\frac{1}{4} \cdot \frac{2}{3}, \text{ or } \frac{2}{12}, \text{ or } \frac{1}{6}$$

$$\frac{1}{5} \cdot \frac{1}{2}, \text{ or } \frac{1}{10}$$

So his tax is $\frac{6}{10}$ of his original inventory $\left(\frac{1}{3}+\frac{1}{6}+\frac{1}{10}\right)$, and that is what 24 is. A simple proportion shows that he started with 40.

Working forward from 40, he pays $\frac{1}{3}$ of 40 as the first tax, leaving himself with $\left(26\,\text{and}\,\frac{2}{3}\right)$. The second tax leaves him with $\frac{3}{4}$ of $\left(26\,\text{and}\,\frac{2}{3}\right)$, which is 20. The third tax is $\frac{1}{5}$ of his remaining 20, or 4. So, he has paid out 24 and he has 16 remaining.

These somewhat complicated calculations may simply be an exercise to train young apprentices for business, but these may also reflect genuine economic conditions in a past age. We simply do not know. But, compared to modern business tax rates, the tax rates in the problem seem quite high and unfair.

If you are inclined to attempt a similar problem, here are two.

PROBLEM 8.3   **A merchant has to give away some of his goods as a tax, as he passes through three territories. At the first stop, he has to part with $\frac{1}{10}$ of his goods; at the second stop, $\frac{1}{9}$ of what remains; and at the third stop, $\frac{1}{8}$. His total tax is 36.**

   What is his original quantity?

PROBLEM 8.4   **A child was not growing and his parents placed him under the care of a yogi who gave the child a regimen of exercise and diet. After a year, the child's height increased by $\frac{1}{7}$. During a second year, the child's height increased by $\frac{1}{8}$, and during a third year, the child's height increased by $\frac{1}{9}$. The grateful parents measured their child and saw the child had attained the height of 60 inches.**

   What was the height before the child met the yogi?

THE WIZARD PROBLEM

Here is a problem that shows the Hindu's expertise with the Pythagorean theorem. If you think about any right triangle, you can convince yourself that (i) the length of the hypotenuse is shorter than the sum of the lengths of the two shorter sides, and (ii) there must be some point along either chosen shorter side where the side can be broken, so that its shorter portion plus the hypotenuse equals its longer portion plus the third side.

In the 3-4-5 triangle, that point is 1 unit from the vertex where the hypotenuse meets the side.

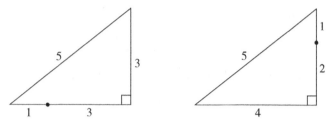

FIGURE 8.1   The perimeter of the 3-4-5 triangle can be broken into two equal lengths.

The break on the left side of Figure 8.1 shows that $5+1=3+3$. The perimeter of the triangle (12) has been broken into two parts, each having a length of 6. The break on the right side of Figure 8.1 shows that $5+1=2+4$. Of course this is a very simple case, but the principle can be extended to any right triangle. The Hindus dressed this fact up in this fanciful problem.

Two monks live on top of a vertical cliff. One descends the cliff and walks on level ground directly to a village. The second monk is a wizard. He flies upward in a straight line, and then diagonally downward to the village. The cliff is 100 yards high. The village is 200 yards away from the base of the cliff. Coincidentally, both monks have traveled the exact same distance when they meet at the village. The question is: **how high did the second monk ascend before he turned to head for the village?**

To solve, we make a right triangle.

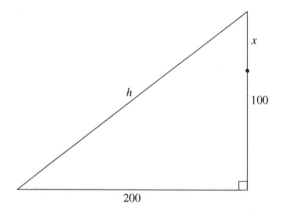

FIGURE 8.2   One monk descends 100 yards and walks 200 yards. The other monk ascends x yards and descends h yards diagonally.

The horizontal leg is simply the 200 yard distance between the base of the cliff and the village. The vertical leg is the sum of the cliff's 100 yard height and x, the distance the monk flew upward. The hypotenuse can be expressed in terms of the Pythagorean theorem, $h^2 = (200)^2 + (100+x)^2$, and we can also say that $h+x=200+100$.

This is enough for us to solve for x. See if you can follow all the steps below, which are expansions, simplifications, and substitutions.

$$h^2 = 40,000 + \left(10,000 + 200x + x^2\right)$$
$$h = 300 - x$$
$$h^2 = 90,000 - 600x + x^2$$
$$40,000 + \left(10,000 + 200x + x^2\right) = 90,000 - 600x + x^2$$
$$50,000 + 200x = 90,000 - 600x$$
$$800x = 40,000$$
$$x = 50$$

So it turns out that the flying wizard went up 50 yards and then diagonally down another 250 yards, matching his fellow monk's more conventional 300 yard journey. The dimensions of the big right triangle, 150-200-250, are the famous 3-4-5 triangle blown up by a factor of 50.

If you are interested in trying similar problems, here are two.

PROBLEM 8.5 **Two monks start on top of a 24-foot tower. One descends the tower and walks 144 feet on level ground directly to the door of a temple. The second monk is a wizard. He flies upward in a straight line, and then diagonally downward to the temple door. Both monks travel the exact same distance.**

How high did the second monk ascend before he turned to head for the temple door?

PROBLEM 8.6 **Two soldiers start at their camp. One walks 5,000 feet directly south and then 15,000 feet directly east. The second soldier walks for some distance directly north, and then for a longer distance in a straight line to the rendezvous point, where he meets the first soldier. Both soldiers walk the exact same distance.**

How far did the second soldier walk before he made his turn to the southeast?

THE SNAKE PROBLEM

Another fine problem involves a spectacular snake that grows at an astonishing rate. In fact, the problem is all about rates. The snake grows at one rate, and it descends into a hole in the ground at a second rate, and we have to figure out how long it takes for the snake to get itself completely underground. Students who have seen the Chinese problem with the cistern with 3 taps may see that the solution to the snake problem depends on a similar logical procedure.

This remarkable snake, 80 feet long when the problem starts, descends $7\frac{1}{2}$ feet into the hole in $\frac{5}{14}$ of a day, and the snake's tail grows $\frac{11}{4}$ feet in $\frac{1}{4}$ of a day.

How long does it take for the ever-growing snake to be fully underground?

The ugly fractions actually cancel out nicely, and the key is determining what happens in 1 day. The descent into the hole, $7\frac{1}{2}$ feet in $\frac{5}{14}$ of a day, becomes

$$\frac{15}{2} \quad \text{in} \quad \frac{5}{14}$$

$$\left(\frac{14}{5}\right)\cdot\left(\frac{15}{2}\right) \quad \text{in} \quad \left(\frac{5}{14}\right)\cdot\left(\frac{14}{5}\right)$$

$$\frac{210}{10} \quad \text{in} \quad \frac{70}{70}$$

$$21\,\text{feet} \quad \text{in} \quad 1\,\text{day}$$

That's how much of the snake goes into the hole every day. Now we have to determine how much the snake grows. The growth, $\frac{11}{4}$ feet in $\frac{1}{4}$ of a day, multiplied by 4, becomes 11 feet in 1 day.

So, in one day, 21 feet of snake disappears into the hole, but another 11 feet of snake appears. It's a net of 10 feet into the hole. Since the snake originally was 80-feet long, the snake is completely underground in $\frac{80}{10} = 8$ days.

You can verify this by thinking of how long the snake is after 8 days: $80\,\text{feet} + (8 \cdot 11)$ feet, or 168 feet. At 21 feet per day, it takes $\frac{168}{21} = 8$ days for the snake to complete its descent.

Two similar problems:

PROBLEM 8.7 **Another remarkable snake, 75-feet long, descends 6 feet into the hole in $\frac{1}{4}$ of a day, and grows 3 feet in $\frac{1}{3}$ of a day.**

How long does it take for this snake to be fully underground?

PROBLEM 8.8 **The poet has already written 800 unread verses but he composes constantly and writes 10 new verses for the lady he is courting every $\frac{1}{5}$ of a day. The lady is astonished and flattered, and she begins to read his entire output at the rate of 25 verses every $\frac{1}{6}$ of a day.**

In how many days will she have read every verse he has written?

THE HORSE PROBLEM

Here is another problem with a set of simultaneous equations. The straightforward way to solve it mimics the basic techniques of matrix algebra, which wasn't invented when this problem was.

> Three friends meet. One owns 7 superior horses, another owns 9 inferior horses, and the third owns 10 camels. They are all pleased to see each other and each immediately gives one of his own animals to each of the others, as a gift. After this exchange, the three friends are coincidentally all equally rich with respect to the value of their animals. Camels are worth $12 apiece.
>
> **What is the value of an inferior horse? What is the value of a superior horse?**

We can approach this problem by keeping a scorecard of who owns which animals and gradually developing an equation that deals with the one price we know, the price of a camel. Call the friends A, B, and C, and tally their animals thus:

	Original			After the swap		
	A	B	C	A	B	C
Superior horses	7	0	0	5	1	1
Inferior horses	0	9	0	1	7	1
Camels	0	0	10	1	1	8

The three friends now have animal collections of equal value, and we can apply the simple algebraic idea of "equals subtracted from equals give equals." Subtract, from each man's collection, one of each type of animal.

	After the subtraction		
	A	B	C
Superior horses	4	0	0
Inferior horses	0	6	0
Camels	0	0	7

Since camels are worth $12 each, 7 are worth $84. This must also be the value of 6 interior horses or 4 superior horses. Thus, inferior horses are worth $14 each, and superior horses are worth $21 each. You can go back to the original problem and apply these prices, and you will see that the scenario has been accurately described.

Here are two additional problems to practice with.

PROBLEM 8.9 **Three friends meet at the village festival. One owns 9 elephants, another owns 18 horses, and the third owns 24 camels. Being friends, each makes a gift to the other two of two of his own animals. Now, they discover to their surprise, they are equally rich (with respect to the value of their animals). If a camel is worth 100 rupees, what is the value of a horse? What is the value of an elephant?**

PROBLEM 8.10   **Four multimillionaire art collectors meet at a social event in New York City. One has 8 Rembrandts, another has 9 Monets, a third has 12 van Goghs, and the fourth has 14 Picassos. These guys like each other so much that they immediately exchange gifts. Each guy gives one of his valuable paintings to each of the other 3. As they're enjoying fine wine and expensive food, they calculate that now, after the exchange, they are coincidentally all equally rich with respect to the value of their paintings. Each Picasso is worth $2,000,000.**

   What is the price of a Rembrandt? What is the price of a Monet? What is the price of a van Gogh?

Let's turn to a different sort of problem, one that is not dressed up in a story, one that illustrates the outstanding analytical ability the ancient Hindu mathematicians had.

   Here is the challenge: to find four numbers whose sum is exactly the same as the sum of their squares.

   If you try to solve this by thinking of ordinary numbers, you immediately see that ordinary numbers will not do. $1+2+3+4 = 10$, while $1+4+9+16 = 30$, so the sum of the squares gets larger too quickly for ordinary numbers to play a role in the solution. In fact, for any number greater than 1, the square of the number will be larger than the number itself, and this forces us to consider fractions less than 1, whose squares are smaller than the fractions themselves.

   Let's arbitrarily pick three small numbers and see what the fourth number would have to be. We are basically saying $(a+b+c+d) = (a^2 +b^2 +c^2 +d^2)$, so if we pick numbers for a, b, and c, that would force a specific value for d. Unfortunately, for many arbitrary values of a, b and c, d would have to be an imaginary number. Let's use a luckier pick for the first three numbers: $\frac{1}{2}, \frac{3}{4}$, and 1. Now,

$$\frac{1}{2}+\frac{3}{4}+1+x=\left(\frac{1}{2}\right)^2 +\left(\frac{3}{4}\right)^2 +\left(1\right)^2 +\left(x\right)^2$$

This is the challenge equation with $a = \frac{1}{2}, b = \frac{3}{4}, c = 1$, and $d = x$. You can work this out and use the quadratic formula to find that x is a complicated number involving the square root of $\frac{11}{4}$. The first step is to collect the terms on the left and square and sum the terms on the right:

$$\frac{9}{4}+x = \frac{29}{16}+x^2$$

This can be expressed as a standard quadratic:

$$x^2 -x-\frac{7}{16} = 0$$

This leads to an answer to the old Hindu riddle, but it is a very unsatisfactory one. It relies on the quadratic formula which, as far as we know, was unknown to the ancient Hindus, and it produces an answer with an irrational number. That answer is

$$x = \frac{1}{2} \pm \frac{1}{4}\sqrt{11}$$

A mathematician named Bhaskara wrote a much more elegant solution, approximately 1000 years ago. Bhaskara proposed and solved many problems, but in many cases we don't know what particular chain of reasoning led to his clever solutions. For this problem, he took a simple sequence of numbers to start: 2, 3, 4, and 5. Then he took the squares of those numbers: 4, 9, 16, and 25. The sum of the four numbers is 14. The sum of the four squares is 54.

He then took the fraction of these sums, $\frac{14}{54}$, and multiplied it by the simple sequence of numbers: $\left[2 \cdot \frac{14}{54}\right], \left[3 \cdot \frac{14}{54}\right], \left[4 \cdot \frac{14}{54}\right]$, and $\left[5 \cdot \frac{14}{54}\right]$.

The resulting numbers were $\frac{28}{54}, \frac{42}{54}, \frac{56}{54}$, and $\frac{70}{54}$.

You should add these last four numbers. You will get 3 and $\frac{17}{27}$.

You should also square these four numbers individually and add the four squares. You will also get 3 and $\frac{17}{27}$. So, the numbers $\frac{28}{54}, \frac{42}{54}, \frac{56}{54}$, and $\frac{70}{54}$, or their reduced forms $\frac{14}{27}, \frac{7}{9}, \frac{28}{27}$, and $\frac{35}{27}$, satisfy the original challenge.

Why did Bhaskara pick 2, 3, 4, and 5 to start his solution? We don't know. But we do know that this is not the only solution with rational fractions. Let's pick another set of numbers in a simple sequence: 3, 5, 7, and 9.

The squares of those numbers are 9, 25, 49, and 81. Duplicating the process above, the sum of the four numbers is 24, and the sum of the four squares is 164.

Multiplying the four numbers by the fraction $\frac{24}{164}$ gives the sequence $\frac{72}{164}, \frac{120}{164}, \frac{168}{164}$, and $\frac{216}{164}$.

You can verify that both the sum of these fractions and the sum of the squares of these fractions is 3 and $\frac{23}{41}$.

PROBLEM 8.11 **Find the number that is simultaneously the sum of the four fractions and the sum of the squares of the four fractions when the starting sequence is 5, 10, 15, 20.**

PROBLEM 8.12   Find the number that is simultaneously the sum of the four fractions and the sum of the squares of the four fractions when the starting sequence is 1, 4, 7, and 10.

EXERCISE

The student is challenged to prove algebraically that the procedure works. It is recommended that the student designates the smallest starting number as x and the difference between numbers as d. That way, the four numbers are $x, x+d, x+2d$, and $x+3d$, and their squares, and the sums of the squares, and the fractions, and the totals can all be expressed in terms of x and d.

The ancient Hindus were very sophisticated calculators of square roots and cube roots, and the last thing we'll look at is an explanation of their remarkably accurate approximation of the square root of 2. We'll get there after a few related digressions.

There were, first of all, some methods they used that were excellent for approximating the square roots of big numbers, like 411, but not particularly accurate when used with small numbers like 2. Here is a formula:

$$\sqrt{n} = \sqrt{A^2 + H} \approx A + \frac{H}{2A} - \frac{\left(\dfrac{H}{2A}\right)^2}{2\left(A + \dfrac{H}{2A}\right)}$$

What are **n**, **H**, and **A** in this formula? The number we are looking to find the square root for is **n**. Let's say we want the square root of 411. 411 is **n**. **A** is the biggest integer whose square is less than **n**. In this case, $A = 20$, because $20^2 = 400$. And **H** is the difference between **n** and A^2, in this case 11. You can work out the formula and see that the ancient Hindu estimate for the square root of 411 is 20.2731 rounded to 4 decimal places, and you can use your calculator to see that this is extremely accurate. To 8 decimal places, the Hindu estimate is 20.27313502 and the calculator answer is 20.27313493. The difference between these two numbers is 0.00000009.

As spectacular as this old formula is for a number like 411, it is pretty pathetic for a number like 2. You might verify that setting $n = 2, A = 1$, and $H = 1$ yields an approximation of 1.25 for the square root of 2. We know the correct value, rounded to 4 decimal places, is 1.4142.

Why the formula is good for big numbers and bad for small numbers is something the student could work out algebraically. But be warned: you will see numbers like A^6 in the denominator of complex fractions.

So we know this formula isn't how the Hindus found their accurate value for the square root of 2, which was: $1 + \dfrac{1}{3} + \dfrac{1}{3 \cdot 4} - \dfrac{1}{3 \cdot 4 \cdot 34}$.

Also of no help was an identity formula involving square roots, credited to the mathematician Bhaskara and called the "Hindu Method," which says

$$\sqrt{a+\sqrt{b}} = \sqrt{a+\sqrt{\dfrac{\left(a^{2}-b\right)}{2}}} + \sqrt{a-\sqrt{\dfrac{\left(a^{2}-b\right)}{2}}}$$

(Curiously, this identity was expressed in an equivalent but convoluted form by Euclid, in the *Elements*.)

To be fair, this formula states an equivalence. It is not designed, like the previous formula, to find square roots. It merely notes an interesting relationship among square roots. And if we put in reasonable numbers for **a** and **b**, like **a** = 7 and **b** = 4, we could verify that the identity works. But, if we were to substitute **a** = 1 and **b** = 1, thinking this might help us find the square root of 2, because the left side would then equal $\sqrt{2}$, then the equation backfires, giving us a nonsensical result. Apparently, **a²** cannot equal **b** in this formula.

A very persuasive explanation of the estimate $1 + \dfrac{1}{3} + \dfrac{1}{3 \cdot 4} - \dfrac{1}{3 \cdot 4 \cdot 34}$ is the geometrical demonstration illustrated in Figure 8.3. A square is drawn on the left side of the figure, and then an identical square is drawn in the middle and cut into pieces. At the right side of Figure 8.3, the pieces of the middle square are arranged around the first square, intending to make a larger square, and this attempt to make a larger square comes tantalizingly close. If the reassembled pieces fit together *exactly* in a second (larger) square, we can see that its area would be twice the area of the original square, and thus the length of its side would be the length of the original square's side multiplied by the square root of 2.

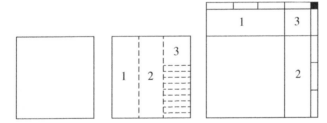

FIGURE 8.3   The Hindu construction for determining the square root of 2.

The dissected square is first cut into thirds vertically (pieces 1 and 2). The third third is cut into one piece (piece 3) that will finish off a square with the original square and pieces 1 and 2, and then 8 smaller pieces that can fit around that square. The resulting figure, not quite a square, has sides that we can express in terms of the length of the side of the original square: $1 + \dfrac{1}{3} + \dfrac{1}{3 \cdot 4}$. It may be more obvious if this is expressed as $1 + \dfrac{1}{3} + \dfrac{1}{12}$. The figure, which is not quite a square, occupies $\dfrac{288}{289}$ of the area of what would be a square with a side length of $1 + \dfrac{1}{3} + \dfrac{1}{12}$. So this number, $1 + \dfrac{1}{3} + \dfrac{1}{12}$, is itself a pretty good estimate for the square root of 2. In decimals, it is 1.41666667, and squaring it yields 2.00694444.

But the ancient Hindus refined this by subtracting the equivalent of $\dfrac{1}{408}$. Where did that peculiar amount come from? Think about filling the darkened corner (at the right side of Fig. 8.3) by slicing off very thin strips of the left side and bottom and chopping them into tiny pieces that can be fitted into the darkened corner. If the original square has a side of 1 and an area of 1, the darkened corner has a side of $\dfrac{1}{12}$ and an area of $\dfrac{1}{144}$. The strip you could slice off the bottom would be a rectangle with a length of $\dfrac{17}{12}$ and an unknown width we can call x. Slicing an equivalent area off the left side and solving for x gives us $x = \dfrac{1}{408}$.

So, the Hindu estimate of $1 + \dfrac{1}{3} + \dfrac{1}{3 \cdot 4} - \dfrac{1}{3 \cdot 4 \cdot 34}$ should be close to perfect. Why isn't it perfect? Did you notice that a very thin slice along the bottom and a very thin slice along the left side have a very tiny overlap in the bottom left corner? That tiny area is counted twice in the calculation. But it is extremely small. The Hindu estimate for the $\sqrt{2}$, $\dfrac{577}{408}$, when squared, gives 2.000006007. This is pretty amazing.

Could we use a similar method to estimate the square root of 3? Why not? In Figure 8.4, three equal squares are drawn, and two are chopped up so that their parts can be arranged around the other one to make a figure that is almost a square.

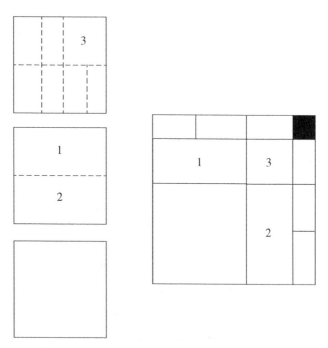

FIGURE 8.4   A geometric approach for determining the square root of 3.

The "almost a square" has sides of $1 + \dfrac{1}{2} + \dfrac{1}{4}$, or $\dfrac{7}{4}$, and $\left(\dfrac{7}{4}\right)^2 = 3.0625$, so already we are very close to the square root of 3. As in the previous calculation for the square root of 2, we have to slice a thin strip off the bottom and off the left side to fill in the missing square on the upper right.

The original square has a side of 1 and an area of 1. The missing corner has an area of $\dfrac{1}{16}$. The thin slice we would take from the bottom would have dimensions $\dfrac{7}{4}$ and x, and we would take the same slice off the left side (making the same "error" of counting the overlap on the bottom left twice). Solving for x, we get $x = \dfrac{1}{56}$, and this makes the formula for the square root of 3: $1 + \dfrac{1}{2} + \dfrac{1}{4} - \dfrac{1}{56}$. This is $\dfrac{97}{56}$.

And $\left(\dfrac{97}{56}\right)^2$ is 3.000318878

EXERCISE

Using the Hindu technique, with the geometrical rearrangement of parts of squares, make an estimate for the square root of 5.

PROBLEM 8.13 Using the Hindu technique, with the formula for A and H, calculate the square root of 250.

PROBLEM 8.14 Using the Hindu technique, with the formula for A and H, calculate the square root of 777.

Suggestions for book or Internet research	
History, archeology	Ancient India
	Early Indus Valley civilization
	Mohenjo Daro
	Maurya Empire, Gupta Empire
Religion, culture	Sanskrit
	Sulvasutra
Mathematical topics	Hindu mathematics
	Hindu numerals
Mathematicians	Bhaskara
	Brahmagupta
	Mahavira

ANSWERS TO PROBLEMS

8.1

Letting P represent pears, and letting B represent bananas, we can make two equations:

$$8P + 5B = 118$$
$$5P + 8B = 103$$

To eliminate the variable B, multiply the first equation by 8 and the second equation by 5

$$64P + 40B = 944$$
$$25P + 40B = 515$$

Subtracting gives

$$39P = 429$$

Dividing by 39 gives

$$P = 11$$

Substituting back into either original equation gives

$$B = 6$$

So, pears cost 11 cents each, and bananas cost 6 cents each.

8.3

At the first stop, the merchant pays $\dfrac{1}{10}$ of his goods as tax, leaving him with $\dfrac{9}{10}$. At the second stop, he gives away $\dfrac{1}{9}$ of his $\dfrac{9}{10}$, leaving him with $\dfrac{8}{10}$. At the third stop, he gives away $\dfrac{1}{8}$ of his $\dfrac{8}{10}$, leaving him with $\dfrac{7}{10}$. His total tax is 36.

But his total tax is also $\dfrac{3}{10}$ of the goods he started with. Since $\dfrac{3}{10}$ is 36, then $\dfrac{1}{10}$ is 12, and his starting total value was 120.

We can check. 120 minus $\left(\dfrac{1}{10} \text{ of } 120\right)$ is 108. 108 minus $\left(\dfrac{1}{9} \text{ of } 108\right)$ is 96. 96 minus $\left(\dfrac{1}{8} \text{ of } 96\right)$ is 84. The original 120 has been reduced by 36 (his total tax).

8.5

It is much easier to compute the distance for the monk who climbed down. He climbed down 24 feet and then walked 144 feet, so his total was 168 feet. The monk who flew must have also traveled 168 feet. Part of that 168 was straight up (x feet), and part was the hypotenuse (h) of a triangle with sides of 144 and $(24 + x)$.

$$\text{So, } h + x = 168, \text{ and } \left(144\right)^2 + \left(24 + x\right)^2 = h^2$$

$$20{,}736 + \left(576 + 48x + x^2\right) = \left(168 - x\right)^2$$

$$\begin{aligned}
21{,}312 + 48x + x^2 &= 28{,}224 - 336x + x^2 \\
21{,}312 + 48x &= 28{,}224 - 336x \\
384x &= 6{,}912 \\
x &= 18
\end{aligned}$$

8.7

Again, the fractions cancel out nicely, and the key is determining what happens in 1 day; 6 feet in $\frac{1}{4}$ of a day becomes 24 feet in 1 day (descent into the hole).

And 3 feet in $\frac{1}{3}$ of a day becomes 9 feet in 1 day (growth of the snake's tail).

So, in one day, 24 feet of snake disappears into the hole, but another 9 feet of snake appears. It's a net of 15 feet. Since the snake originally was 75 feet long, the snake is completely underground in $\frac{75}{15} = 5$ days.

8.9

Again, we make a scorecard of who owns which animals and gradually develop an equation that deals with the one price we know, the price of a camel. Remember that each man gives away 4 animals, two to each of two friends.

	Original			After the swap		
	A	B	C	A	B	C
Elephants	9	0	0	5	2	2
Horses	0	18	0	2	14	2
Camels	0	0	24	2	2	20

The three friends now have animal collections of equal value, and we can subtract two of each type of animal.

	After the subtraction		
	A	B	C
Elephants	3	0	0
Horses	0	12	0
Camels	0	0	18

Since camels are worth 100 rupees each, 18 are worth 1800 rupees. This must also be the value of 12 horses or 3 elephants. Thus, horses are worth 150 rupees, and elephants are worth 600 rupees.

8.11

Starting with 5, 10, 15, 20, the sum is 50. The squares of these numbers are: 25, 100, 225, and 400. These squares add up to 750. So the ratio of the sums is $\frac{50}{750}$ or $\frac{1}{15}$. We multiply the starting numbers by $\frac{1}{15}$ and get $\frac{5}{15}, \frac{10}{15}, \frac{15}{15}$, and $\frac{20}{15}$, which reduce to $\frac{1}{3}, \frac{2}{3}, 1$, and $\frac{4}{3}$. The sum of these numbers is $3\frac{1}{3}$, and you can verify that the squares of these numbers $\left(\frac{1}{9}, \frac{4}{9}, 1, \text{ and } \frac{16}{9}\right)$ also add up to $3\frac{1}{3}$.

8.13

$\sqrt{250} = \sqrt{225 + 25} = \sqrt{15^2 + 25}$, so we put $A = 15$ and $H = 25$ into the formula.

This gives $15 + \dfrac{25}{30} - \dfrac{\left(\dfrac{25}{30}\right)^2}{2 \cdot \left(15 + \dfrac{25}{30}\right)}$

This is 15.81140351

Squaring 15.81140351 gives 250.0004809

The calculator gives 15.8113883 for the square root of 250, so once again the Hindu formula is extremely accurate.

9

EARLY ARABIAN MATHEMATICS

From roughly AD 500 to AD 1300, hundreds of mathematicians worked in the area we now call the Middle East. We know their names because many of their books have endured through the subsequent centuries. They wrote in Arabic. They were supported by rulers who valued their contributions to astronomy, architecture, and commerce. Their quest for theoretical, less practical knowledge was also admired and appreciated.

We tend to lump all the work of these mathematicians together under the name Arabian mathematics, but some were Persian, some were Turks, some lived in North Africa, and some lived in areas controlled at various times by the Islamic caliphate, like Spain. So another reasonable name for Arabian mathematics is Islamic mathematics.

There is a real controversy about what was the Arabs' main contribution to math. They preserved and translated the work of the ancient Greeks, and they introduced the Hindu numerals to Europe. Remember: Europe was in what we call the *Dark Ages* during this time, where scholarship had declined and the collapse of the Roman Empire led to a chaotic, fragmented, and insecure society. The European Renaissance was propelled by ancient wisdom, translated from Greek to Arabic and then from Arabic to Latin so that it could be disseminated to the churchmen and scholars in university towns in Italy, Germany, and France. So, one view is that the Arabs chiefly preserved Greek knowledge.

But the Arab writers also assessed and extended the Greek heritage. They did new work in geometry and algebra, and a second view is that the massive accumulation of all these new ideas is the real contribution. Arabian mathematicians went beyond what we can firmly attribute to their Greek (or even Hindu)

An Introduction to the Early Development of Mathematics, First Edition. Michael K. J. Goodman.
© 2016 John Wiley & Sons, Inc. Published 2016 by John Wiley & Sons, Inc.

predecessors in conic sections, spherical geometry, polynomials, 5th roots of equations, and tessellations.

Sometimes it is hard to tell what is a genuinely new contribution and what is an improvement or a rediscovery. But ultimately, to argue is pointless; it is clear that the Arabian mathematicians both transmitted and created.

Not very many of the Arabian/Islamic mathematicians are widely known by name among American students today, but one name that is generally recognized is that of Omar Khayyam. He is widely known because of a long poem he wrote, the *Rubaiyat*, and students are often surprised that this romantic poet was also a mathematical genius. He worked in the 11th century, in Persia, and his most remarkable result was a method for solving cubic equations by drawing circles and hyperbolas.

A cubic equation, in our modern notation, has the general form $ax^3 + bx^2 + cx + d = 0$, where a, b, c, and d are real numbers.

High school students are familiar with the considerably easier general form of the quadratic equation, $ax^2 + bx + c = 0$, which is solved by the quadratic formula

$$x = \frac{-b \pm \sqrt{b^2 - 4ac}}{2a}$$

A similar "magic" formula for solving cubic equations was a goal of ancient and medieval mathematicians, and eventually early in the Renaissance Italian algebra experts were finally able to find a technique, which even today strikes us as complicated and clunky, illustrating just how hard the problem is.

Of course, if we know the solution in advance, or have an easily factored expression, finding the roots of a cubic is not hard, but most cubic equations are very hard.

For example, $x^3 - 7x^2 + 7x + 15 = 0$ is easily solved because only a little tinkering reveals the factors

$$(x+1)(x-3)(x-5) = 0$$

A scholar in the 11th century would have happily seen $x=3$ and $x=5$ as solutions, but would have rejected $x=-1$ because negative numbers were regarded with great suspicion and generally not recognized as legitimate numbers at all. The arithmetical check of the two "good" solutions is as follows:

$$(3)^3 - 7(3)^2 + 7(3) + 15 = 27 - 63 + 21 + 15 = 0$$
$$(5)^3 - 7(5)^2 + 7(5) + 15 = 125 - 175 + 35 + 15 = 0$$

The equation $x^3 - 7x^2 + 7x + 15 = 0$ can be solved because it is easily factored. But thousands of similar equations cannot be solved because they are not factorable. These equations include

$$x^3 - 7x^2 + 7x + 16 = 0$$
$$x^3 - 7x^2 + 7x + 17 = 0$$
$$x^3 - 7x^2 + 7x + 18 = 0$$

and, in general, any minor change to one of the coefficients of the original equation creates a new equation that cannot be readily solved.

If you have a graphing calculator, you could verify this by putting

$$y = x^3 - 7x^2 + 7x + 15$$

into the function or formula space and observing how nicely the up-and-down curve goes through the points $(-1, 0)$, $(3, 0)$, and $(5, 0)$. When the 15 at the end is replaced by 16, 17, or 18, the up-and-down curve shifts slightly, no longer meeting the x-axis at precise integer values.

Omar Khayyam's method for solving cubics involved algebra and geometry. He classified cubic equations into more than a dozen varieties, depending on whether the x^3 term was accompanied by an x^2 term, or an x term, and how these terms were connected. He would distinguish between $x^3 + 12 = 19x$, $2x^3 = 8x^2$, and $x^3 + 5x^2 = 144$, for example, and while the distinctions may seem fussy to us, they made a big difference to him because he varied his solving technique based on the classification. His general approach was to draw two conic sections (circles, ellipses, parabolas, and hyperbolas), but which conic sections he drew depended on which variety of cubic equation he had. We can see easily that $x = 4$ is a solution all three of the cubics mentioned earlier, but Omar Khayyam would have found $x = 4$ by three different constructions.

He was a master at drawing the conics and finding their intersections, and he was a master at drawing all the little line segments that related the coefficients of the cubic equation to the coefficients of the conics. He could draw one line that was in proportion to a second line exactly as a third line was in proportion to fourth. He could construct a line parallel to another, or perpendicular to another, or equal in length to another. He could find a diameter for a circle or an asymptote for a hyperbola. He could find one line segment proportional to the square or cube of another.

He was, undoubtedly, gifted, and the whole idea of transforming a cubic equation to a combination of conic sections is remarkable, but it's important to see this remarkable cleverness in context. Omar Khayyam did his work 400 years after the first Arab mathematicians began translating the works of the Greek geniuses, especially Euclid's *Elements*. Around AD 860, a mathematician named al-Mahani worked extensively with spheres and analyzed the general equation $x^3 + a^2b = ax^2$. At the same time, another scholar, al-Himsi, made a new translation of Apollonius' *Conics*. A century before Omar Khayyam was born, another mathematician, Ibrahim ibn Sinan, wrote a book describing precise techniques for constructing ellipses, parabolas, and hyperbolas. Every element of Omar's discovery was more or less available and waiting for someone like him to put it all together. Al-Haytham, who died just before Omar Khayyam was born, played with this algebraic nugget: if four numbers are in proportion such that $\dfrac{b}{c} = \dfrac{c}{d} = \dfrac{d}{a}$, then $\dfrac{b}{c} \cdot \dfrac{b}{c} = \dfrac{c}{d} \cdot \dfrac{d}{a}$, which means that $\dfrac{b^2}{c^2} = \dfrac{c}{a}$. Crossmultiplying yields $c^3 = b^2 a$, where $c = \sqrt[3]{b^2 a}$ and if b happens to be exactly equal to 1, then c is exactly the cube root of a.

Omar Khayyam synthesized all the ideas available to him and took the study of cubic equations further than anyone else ever had. And, over a long career in mathematics, he did much more, working on problems in arithmetic, algebra, and geometry, and contributing to advances in a variety of topics, including polynomials, irrational numbers, the binomial theorem, and parallel lines.

The Arabs had a general fascination with geometric figures. Partly this had to do with their art which, influenced by their religion, emphasized abstract patterns over realistic representations. Partly, it had to do with their elaborate study of classical Greek mathematics, which as we have seen had a strong bias toward geometry. Here is a typical clever Arabian extension of a Greek idea: any triangle can be marked so that the sum of the squares of two sides equals the sum of two rectangles drawn along the third side.

This sounds something like the Pythagorean theorem, and that is where we should start in our investigation of this idea. The simplest right triangle with sides that are all integers is the 3-4-5 triangle. This small triangle can be enlarged by multiplying each side by a constant, and therefore we can visualize right triangles with sides of 6-8-10, 9-12-15, 12-16-20, and 15-20-25.

The last three of these can fit together quite nicely, as in Figure 9.1 where CH is the altitude on AB, in triangle ABC. ABC is the big triangle, with the right angle at C: $(BC)^2 + (AC)^2 = (AB)^2$, or $15^2 + 20^2 = 25^2$. The altitude CH splits ABC into two smaller, similar triangles, AHC and BHC, with the right angles at H. In AHC, HC is the shortest side (by length) and $12^2 + 16^2 = 20^2$. In BHC, HC is the middle side (by length) and $9^2 + 12^2 = 15^2$.

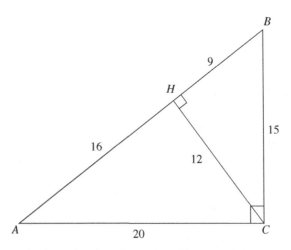

FIGURE 9.1 The 9-12-15 triangle and the 12-16-20 triangle joined to make a 15-20-25 triangle.

Drawing the squares on the sides of triangle ABC gives us the obvious Pythagorean relationship $225 + 400 = 625$, but extending the altitude CH into the square of the hypotenuse splits that square into two interesting parts, as shown in

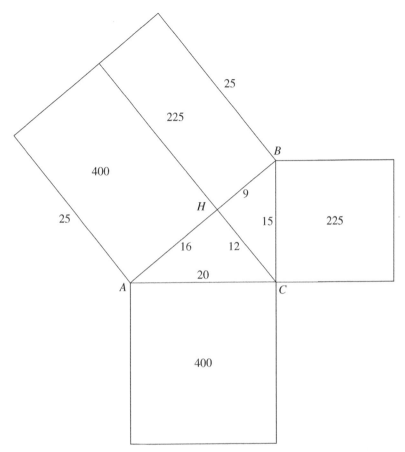

FIGURE 9.2 The square of the hypotenuse of the 15-20-25 triangle has been split into two pieces. The area of each piece exactly matches the area of the square of one of the shorter sides.

Figure 9.2. 9×25 matches the square of the smallest side (225), and 16×25 matches the square of the middle side (400).

A key point here is that we are looking at three similar triangles, *similar* meaning that the triangles, although they have different sizes, have exactly the same angles, which causes the ratio of the lengths of the sides to be the same from triangle to triangle. The angle at A (we could call it $\angle BAC$) equals $\angle HCB$; and the angle at B (we could call it $\angle ABC$) equals $\angle HCA$; and the angle at C is a right angle, as are $\angle AHC$ and $\angle BHC$.

In the 9th century, Thabit ibn Qurra, who was born in Turkey and educated in Baghdad, had the profound insight that $\triangle ABC$ didn't have to be a right triangle and line CH in the diagram could be replaced by two lines. In Figure 9.3, the angles at X and Y equal the angle at C (or particularly, $\angle ACB = \angle AXC = \angle BYC$) and triangles ABC, ACX, and BCY are similar. Thabit ibn Qurra asserted that

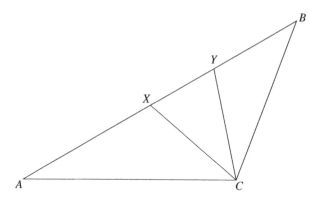

FIGURE 9.3   In triangle *ABC*, *X* and *Y* have been placed so that triangles *AXC* and *BYC* are similar to *ABC*. Now $(AC)^2+(BC)^2=(AB)(AX+BY)$.

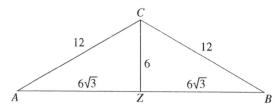

FIGURE 9.4   Two 30°-60°-90° triangles have been joined to form a 30°-30°-120° triangle.

$(AC)^2+(BC)^2=(AB)(AX+BY)$, and this is obviously similar to the Pythagorean theorem. It says that the square of one side of a triangle, plus the square of a second side, is equal to something that is not quite the square of the third side, but instead the third side multiplied by a specific part of itself.

This is a pretty peculiar idea, and it's certainly not obvious. Let's make it easy on ourselves and test the calculations out with a triangle that has angles and sides that are easy to compute with. It is well known that the sides of a 30°-60°-90° triangle are in the ratio: $x, x\sqrt{3}$, and $2x$. The short side, x, is opposite the 30° angle; the long side, $2x$, is opposite the 90° angle; and the intermediate side, $x\sqrt{3}$, is opposite the 60° angle.

In Figure 9.4, two 30°-60°-90° triangles, $\triangle AZC$ and $\triangle BZC$, with lengths 6, $6\sqrt{3}$, and 12, have been joined at their right angles to make $\triangle ABC$, which therefore has angles of 30°, 30°, and 120°. The length of *AC* is 12, and the length of *CB* is 12. The length of *AB* is $12\sqrt{3}$. This is a nice, symmetric, isosceles triangle.

We can apply Thabit ibn Qurra's idea by constructing X and Y so that $\angle AXC = \angle BYC = 120°$ (Fig. 9.5). Now, $\triangle AXC$ and $\triangle BYC$ are isosceles triangles, and $\triangle XCY$ is equilateral, and *AB* is split into three segments, *AX*, *XY*, and *YB*, which are all $4\sqrt{3}$ in length. Rectangles are constructed along *AX* and *BY*, so that the lengths of these rectangles are also $12\sqrt{3}$, like side *AB*. Squares are constructed along *AC* and *BC*.

It's no great shock that $12 \times 12 = 144$ and $4\sqrt{3} \times 12\sqrt{3} = 144$, but it's interesting that each square has a rectangular counterpart that is equal in area.

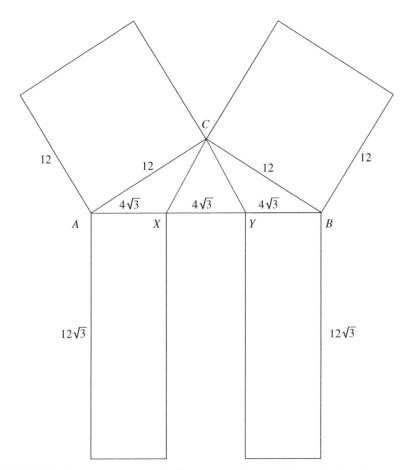

FIGURE 9.5 Squares and rectangles have been added, and their calculated areas match Thabit ibn Qurra's prediction.

In Figure 9.6 we see an unsymmetrical triangle, commonly called a scalene triangle. Its sides have lengths of 8, 10, and 15. It is obviously not a right triangle $(8^2 + 10^2 \neq 15^2)$, and in fact if we know and use the law of cosines, we can determine that the angles are approximately $29\frac{1}{2}°$, $38°$, and $112\frac{1}{2}°$. (It's interesting to know these measurements, but not essential for our purposes.) Now X and Y have been placed on AB so that the angles at X and Y are the same as the angle at C, close to $112\frac{1}{2}°$.

Thabit ibn Qurra's assertion is that $(AC)^2 + (BC)^2 = (AB)(AX + BY)$, and we can use the ratio property of similar triangles to compute that $(AX) = \frac{100}{15}$ and $(BY) = \frac{64}{15}$. If we build the rectangles, as in Figure 9.7, we see again that the squares and rectangles are counterparts by area, and Thabit ibn Qurra's formula works.

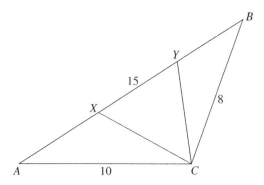

FIGURE 9.6   Smaller scalene triangles are constructed within a larger scalene triangle, to test Thabit ibn Qurra's theorem.

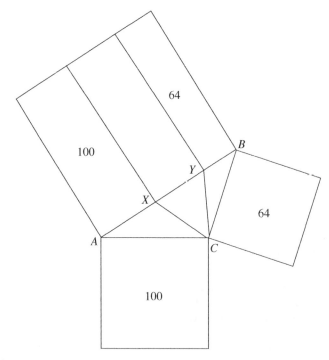

FIGURE 9.7   The sum of the areas of the rectangles again matches the sum of the areas of the squares.

PROBLEM 9.1   Given a triangle ABC with sides of 7, 11, and 16, with AB the long side, what are the lengths of AX and BY when X and Y are drawn on AB so that $\angle X = \angle Y = \angle C$?

PROBLEM 9.2   Given a triangle ABC with sides of 12, 15, and 21, with AB the long side, what are the lengths of AX and BY when X and Y are drawn on AB so that $\angle X = \angle Y = \angle C$?

Thabit ibn Qurra had ideas in arithmetic as well as in geometry. You may recall that the Greeks had a special interest in the whole number divisors of numbers and that Pythagoras had found that the divisors of two numbers, 220 and 284, had the curious property that the sum of the divisors of 220 was 284, and the sum of the divisors of 284 was 220, making 220 and 284 so-called amicable numbers. Thabit ibn Qurra saw that $220 = 4 \times 5 \times 11$ and that $284 = 4 \times 71$, and that 5, 11, and 71 were prime numbers that could be related to each other via a formula.

$$5 \text{ is } (3 \times 2) - 1$$

$$11 \text{ is } (3 \times 4) - 1$$

$$71 \text{ is } (3 \times 3 \times 2 \times 4) - 1$$

$$\text{Notice that } (5 \times 11) + (5 + 11) = 71.$$

Thabit ibn Qurra proposed a general algebraic rule for finding amicable numbers based on the patterns above. His rule depends on substituting a value for **n** into these 5 expressions:

The smallest prime, **p**, has to be $3 \cdot 2^{n-1} - 1$.

The next prime, **q**, has to be $3 \cdot 2^n - 1$.

And the largest prime, **r**, has to be $9 \cdot 2^{2n-1} - 1$.

The amicable numbers are then: $2^n \cdot \mathbf{p} \cdot \mathbf{q}$ and $2^n \cdot \mathbf{r}$.

EXERCISE

Show that the amicable pair 220 and 284 comes from putting 2 in for *n* in Thabit ibn Qurra's formulas.

If Thabit ibn Qurra ever found another pair of amicable numbers using his rule, all records of it are lost. But the rule also works for the next pair of amicable numbers, which was discovered in the 17th century.

$$23 \text{ is } (3 \times 8) - 1$$

$$47 \text{ is } (3 \times 16) - 1$$

$$1151 \text{ is } (3 \times 3 \times 8 \times 16) - 1$$

You may want to calculate and confirm that $(23 \times 47) + (23 + 47) = 1151$.

$16 \times 23 \times 47 = 17{,}296$ and $16 \times 1151 = 18{,}416$ and these two 5-digit numbers are also amicable numbers.

EXERCISE

Show that the amicable pair 17,296 and 18,416 comes from putting 4 in for *n* in Thabit ibn Qurra's formulas.

PROBLEM 9.3 Show that $n = 3$ doesn't lead to an amicable pair because the formulas for *p*, *q*, and *r* do not produce three prime numbers.

PROBLEM 9.4 **Show that $n=5$ doesn't lead to an amicable pair because the formulas for p, q, and r do not produce three prime numbers.**

In the 8th century, there was a ban on portraying any human figure in religious art. This led to the development of spectacular decorations, featuring curves and polygons and intricate networks of lines. A lot of this work focused on the task of filling spaces completely with symmetrical interlocking shapes, what we now call tiling or tessellations. The number of possible symmetrical patterns is large, but it appears that every interesting, attractive design that fully covers a flat surface by using only regular repeating polygons was known to or discovered by the geometric specialists among the Arabian mathematicians.

It is easy to demonstrate that squares and rectangles alone can completely fill a flat area, because their 90° angles conveniently converge at the points where the squares or rectangles meet. The squares and rectangles can even be shifted horizontally within a row and, as long as the area to be filled is thought of as extending forever, they still can completely fill the flat area. Quadrilaterals that can be made by distorting squares and rectangles, like rhombuses and parallelograms, can also fill a flat area. Triangles of any shape can fill a flat area, and equilateral triangles particularly make patterns that can please or trick the eye. Regular hexagons can fill a flat area. Regular pentagons however cannot, and this appears to have inspired some Arabian geometry with pentagons that just miss being regular.

A regular pentagon has five equal angles (each 108°) and five equal sides (any length, as long as all five are the same).

EXERCISE

Demonstrate why the interior angles of a regular pentagon are 108°

Two very clever Arabian mathematicians, Abu Kamil and Abu Sahl, gave some thought to the problem of putting a square around a pentagon, so that all five of the pentagon's vertexes touched the sides of the square. Ideally, they would have liked to put a square around a regular pentagon, but that turned out to be impossible: the regular pentagon's width exceeds its height. Let's see why that is.

In Figure 9.8, regular pentagon $ABCDE$ is inscribed in rectangle $FGHJ$. EB, the width of the pentagon, is equal to JH, the width of the rectangle. AK, the height of the pentagon, is equal to GH, the height of the rectangle. If JH and AK were equal, the rectangle would be a square, but they are not equal. We could simply measure them to verify that they are not equal. If the sides of the pentagon (AB, BC, etc.) were each 1, we would measure that JH was about 1.618 and GH was about 1.539.

The Arabs did better than this. They knew exactly how long JH was because Euclid had already calculated the length of the diagonal of a pentagon. They found GH by connecting A to D and calculating AK by the Pythagorean theorem. We can see, in Figure 9.8, that AD would be the hypotenuse of right triangle AKD, and we know $DK = \dfrac{1}{2}$ and $AD = \dfrac{1+\sqrt{5}}{2}$. This leads to an exact value for AK of $\dfrac{\sqrt{5+2\sqrt{5}}}{2}$.

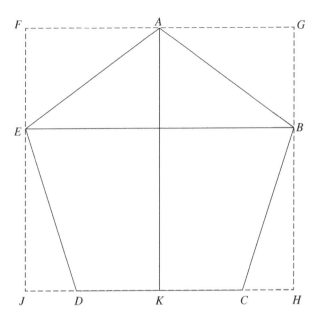

FIGURE 9.8 The rectangle drawn around a regular pentagon is not a square.

Since the regular pentagon is a little too fat to fit into a square, the Arab mathematicians worked with the next best thing: a pentagon with five equal sides and some (but not five) equal angles, which is called an equilateral pentagon. They found two ways to stick an equilateral pentagon inside a square, and a natural question arose: what is the length of the side of the surrounding square in comparison to the side of the equilateral pentagon? Answering this question involved some very careful algebra.

Around AD 900 Abu Kamil worked with the configuration shown in Figure 9.9, wedging two sides of the pentagon (*AE* and *DE*) into one 90° corner of the square and letting the angle at *A* match the angle at *D*, and the angle at *B* match the angle at *C*.

The natural (but challenging) question is: how long is *EF* if *EA* is 1?

We can call the length of a side of the square **s** and the length of a side of the inscribed pentagon 1. So, $AB = BC = CD = DE = EA = 1$ and $EF = FG = GH = HE = s$.

Since *BGC* and *AFB* are right triangles, we can use the Pythagorean theorem to find the lengths of their sides, and this will lead to finding **s**, the side of the square.

BGC turns out to be a 45°-45°-90° triangle, and since its hypotenuse is 1, its smaller sides are $\dfrac{1}{\sqrt{2}}$, which we can rationalize to $\dfrac{\sqrt{2}}{2}$. *BG* is $\dfrac{\sqrt{2}}{2}$.

This is useful because *FG* is a side of the square, $FG = s$, and $FG = FB + BG$, and we can therefore say $FB = FG - BG$, or $FB = s - \dfrac{\sqrt{2}}{2}$.

Similarly, $EF = s$, and $EF = EA + AF$, so $AF = EF - EA$, or $AF = (s - 1)$.

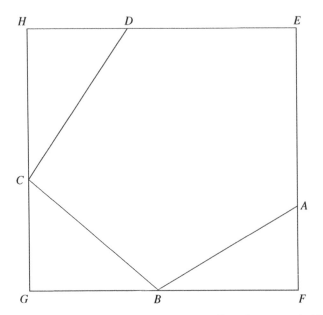

FIGURE 9.9   Abu Kamil found one way to put an equilateral pentagon inside a square.

Using the familiar $a^2 + b^2 = c^2$ on triangle AFB, we can say

$$\left(s - \frac{\sqrt{2}}{2}\right)^2 + (s-1)^2 = 1^2$$

This is an equation with only one unknown: s. So, in principle, we can solve it. Getting rid of all the radicals leads to a 4th-degree equation, and solving a *quartic* equation (with x^4 terms) is even more difficult than solving Omar Khayyam's *cubic* equations (with x^3 terms). Nonetheless, Abu Kamil solved it.

The quartic equation is $4s^4 - 8s^3 + 4s^2 - 2s + \frac{1}{2} = 0$.

The length of the side of the square, s, turns out to be around 1.63.

About a hundred years later, Abu Sahl worked on the pentagon-within-the-square problem, using a different configuration, shown in Figure 9.10. His equilateral pentagon has only one side lying on a side of the square (AE on JM), and the angle at A matches the angle at E, while the angle at B matches the angle at D. C sits at the midpoint of KL.

It turns out that the square in Abu Sahl's arrangement isn't exactly the same size as the square in Abu Kamil's arrangement. (Another way to say this is: if we tried to put Abu Sahl's pentagon and Abu Kamil's pentagon in the same square, Abu Sahl's would be slightly smaller.)

In Abu Sahl's arrangement, let's call the length of a side of the square r (and the length of a side of the inscribed pentagon 1).

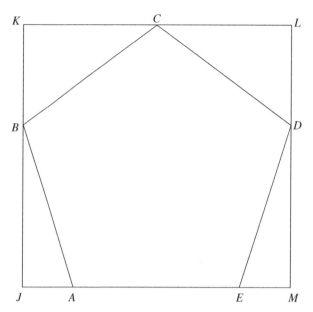

FIGURE 9.10 Abu Sahl found another way to put an equilateral pentagon inside a square.

So, $AB = BC = CD = DE = EA = 1$, and $JK = KL = LM = MJ = r$.
Again, we can work with right triangles.

EXERCISE

Show that CK is $\dfrac{r}{2}$ and use the Pythagorean theorem to find BK.

Show that AJ is $\dfrac{r-1}{2}$ and use the Pythagorean theorem to find JB.

Use BK and JB to express JK in terms of r.

Another pile of arithmetic and another 4th-degree equation tell us that r in the Abu Sahl arrangement is approximately 1.57.

We might well ask why Abu Kamil and Abu Sahl exerted themselves so much, why they were so fascinated with the exact calculation of the ratios of the pentagon and the square. Surely, there is no practical reason to calculate such an obscure thing. The probable answer is: they did it because it was such an enormous challenge. They did it because they were interested in knowledge for its own sake, and they belonged to a community of scholars who kept trying to increase the sum of all mathematical knowledge, regardless of whether each additional fact had a practical use or not.

But among their contemporaries were people who studied the night sky with an extremely practical objective. These mathematicians made innumerable astronomical observations and pioneered the study of the geometry of spheres because they had to resolve an extremely important religious question.

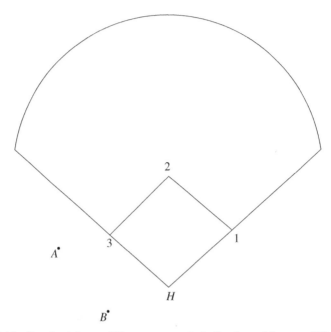

FIGURE 9.11 People sitting at different seats at the ballpark would turn at different angles to see home plate.

One of the daily obligations of a Muslim is to pray, facing Mecca. This makes it important to know, wherever you are, where Mecca is, and so you need north–south and east–west clues that are available from precise observation of the sun and the stars. The learned man who could tell the devout which wall of the mosque to face, and indeed where to build the mosque, was a well respected person.

Let's consider this geographical problem by thinking about a simpler, modern situation. Imagine that you are at a major league baseball stadium with tens of thousands of seats. Depending on where your seat is, when you look forward you are looking directly at home plate, third base, center field, first base, or some other part of the field. To observe any other part of the field, you must turn your eyes or your head at various angles. In Figure 9.11, a person sitting in seat A would say a straight line from her seat to first base passes directly through third base. A person sitting in seat B might find that he had to turn his head 60° to look alternately at first base and third base.

Now, both spectators know, because of their familiarity with baseball fields, where home plate is, relative to their observations of first base and third base. The man in seat B would know to face first base and glance slightly to his right to locate home plate. The woman in seat A would know that she also had to look to the right of the first base, but for her it would be at a much greater angle.

This is essentially what early Arabian mathematicians had to do, but on a much bigger scale (thousands of miles instead of hundreds of feet), and in three dimensions instead of two, and with moving landmarks like the sun and the moon instead of stationary ones like third base. Rather remarkably, they generally did it.

There is an interesting complication that you see if you have access simultaneously to a world atlas and a globe. Pick out New York City and Mecca on the largest flat map in the atlas. The approximate straight line joining New York and Mecca runs mostly east and slightly south. Now take the globe, and starting in New York trace a line northeast, slightly to the side of Nova Scotia. Remarkably, as you trace this line onward, along an arc, you also come to Mecca. If you have selected the right arc, it will eventually take you all the way around the world and back to New York, covering the globe in the same way that the equator circles all around it or that any line of longitude joins the north and south poles. These huge circles are called great circles; they are the largest circles that can be traced on the surface of a sphere.

One of the earliest and most important names in Arabian mathematics is Al-Khwarizmi. In the 9th century, he was the director of a Baghdad institution called The House of Wisdom, which combined the functions of a library and a university. Al-Khwarizmi oversaw the translation of many Greek books, and he wrote his own book of math problems, explaining the methods of algebra. He is credited with both coining the word 'algebra' and solidifying the use of Hindu–Arabic numerals.

There has been some criticism of his book. Some say not enough of it is original; it is translated or derived from earlier Greek or Hindu texts. Some say it focuses too heavily on problems involving the division of estates, a practical (instead of theoretical) emphasis. Some say the mathematics itself is unsophisticated, compared to the base of knowledge established by the Greeks or even the Babylonians. But the book did accomplish some very important things: he classified varieties of equations and explained clear standardized methods for solving them. This idea, that a set of rules could be used again and again, was not quite revolutionary at the time but certainly a new and better approach to the needs of teaching and research.

Al-Khwarizmi developed the method for solving quadratic equations by completing the square, a geometrical approach. Let's say we wanted to solve the general equation

$$\mathbf{x}^2 + \mathbf{b}\mathbf{x} = \mathbf{c}$$

You probably have been taught to take half of $\mathbf{b}$, and square it, and add it to both sides of the equation, giving

$$\mathbf{x}^2 + \mathbf{b}\mathbf{x} + \left(\frac{\mathbf{b}}{2}\right)^2 = \mathbf{c} + \left(\frac{\mathbf{b}}{2}\right)^2$$

Now, what's on the left side of the equation is a perfect square, $\left(\mathbf{x} + \frac{\mathbf{b}}{2}\right)^2$, and what's on the right side is a number, since $\mathbf{b}$ and $\mathbf{c}$ are numbers. You can take the square root of both sides and easily find $\mathbf{x}$.

Since this equation, $x^2 + bx = c$, looks so general and abstract, let's modify it by putting in some numbers for **b** and **c** that lead to an easy solution:

$$x^2 + 20x = 341$$

The place to start is the $20x$. We take half of the 20, which is 10, and square it, which gives 100. We add 100 to both sides of the equation.

$$x^2 + 20x + 100 = 441$$

The left side of the equation is now a perfect square. It is $(x + 10)^2$. Coincidentally, because we rigged the original equation with 341, the number on the right side is also a perfect square: 441 is $(21)^2$.

So we can take the square root of both sides of the equation and get

$$(x + 10) = 21$$

which leads to the obvious solution $x = 11$.

To make sure that we have played by the rules and done nothing underhanded, we can return to the original equation, $x^2 + 20x = 341$, and evaluate it, substituting 11 for **x**:

We get $121 + 220 = 341$, and this is a true statement, verifying that **x** is a solution.

Al-Khwarizmi showed this process as a geometrical construction. Starting with the expression $x^2 + 20x$, he said x^2 was a square and $20x$ was a rectangle, and he broke the rectangle into four equal pieces and arranged them around the square, as shown in Figure 9.12. The square in the middle clearly has sides of length x and an area of x^2. The four rectangles around the square have sides of **x** and 5, and each thus has an area of **5x**, making the combined area of all four **20x**. This $x^2 + 20x$ shape, according to the original equation, had an area of 341.

The 12-sided shape in Figure 9.12 has no special beauty—we might say it looks like an "incomplete square." But Al-Khwarizmi made it more beautiful by adding to its corners, completing the square. Because the rectangles have sides of **x** and 5, and the **x**'s are aligned along the central square, the squares that he added in Figure 9.13 were 5-by-5, each with an area of 25. Adding these four squares added 100 (4 times 25) to the previous area of 341. Adding these four squares "completed the square," forming a square, which we can now see has sides of length **x + 10**. It is easy to see that $(x + 10)^2$ in this case must be 441, leading to the conclusion we already know: $x = 11$.

PROBLEM 9.5   Use the technique of completing the square, algebraically and geometrically, to solve $x^2 + 28x = 204$.

PROBLEM 9.6   Use the technique of completing the square, algebraically and geometrically, to solve $x^2 + 10x = 1739$.

Al-Khwarizmi had a lot to say about equations, especially quadratic equations, which is not surprising, given how clever and foolproof his completing the square technique was. He classified equations and described methods to solve each type. To our

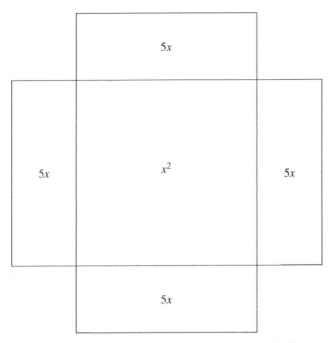

FIGURE 9.12   The geometrical representation of $x^2 + 20x$.

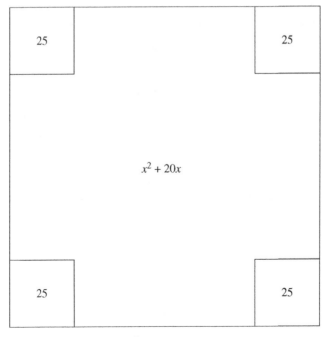

FIGURE 9.13   Completing the square.

modern eyes, his classification appears overly complicated, but that's because he, like his contemporaries, did not trust or use negative numbers, and this forced him to express similar equations as if they were different. Also, it is worth pointing out, he used words, not symbols, to describe his equations. The following description, employing symbols, is a modernization of Al-Khwarizmi's classification.

The first and easiest type of equation to solve is $\mathbf{ax} = \mathbf{b}$, where we are told $\mathbf{a}$ and $\mathbf{b}$ and have to determine $\mathbf{x}$. We use this formulation today, and we call it a linear equation because its graph is a straight line. A key consideration is that $\mathbf{x}$ has no exponent, so there are no squares involved. There is however a common convention now to express equations so that they state that an algebraic quantity equals zero, so $\mathbf{ax} = \mathbf{b}$ could be rewritten as $\mathbf{ax} - \mathbf{b} = \mathbf{0}$.

Today, we use a general expression for all quadratic equations, the previously mentioned $\mathbf{ax^2} + \mathbf{bx} + \mathbf{c} = \mathbf{0}$. The graphs of these equations are always parabolas, and we can solve these equations by completing the square or factoring or by other methods. But let's consider those three terms, $\mathbf{ax^2}$, $\mathbf{bx}$, and $\mathbf{c}$: if all three are positive, then they cannot add up to zero. Either at least one of the coefficients ($\mathbf{a}$, $\mathbf{b}$, and $\mathbf{c}$) must be negative, or $\mathbf{x}$ itself must be negative. This doesn't bother us, but in Al-Khwarizmi's time no one was comfortable with negative numbers. This led Al-Khwarizmi to classify five distinct types of quadratic equations. We can look at the types quickly in a theoretical way and then use actual numbers to make them more recognizable.

One type is $\mathbf{ax^2} = \mathbf{b}$. This says a number, $\mathbf{a}$, times a square equals another number, $\mathbf{b}$. This formulation is very similar to Al-Khwarizmi's generalization of a linear equation, the only difference being the exponent applied to $\mathbf{x}$.

A second type is $\mathbf{ax^2} = \mathbf{bx}$. This says a number, $\mathbf{a}$, times a square equals another number, $\mathbf{b}$, times the square root of the square.

The three other types involve adding a $\mathbf{c}$ term to this second type:

$$ax^2 + bx = c$$
$$ax^2 + c = bx$$
$$ax^2 = bx + c$$

We might summarize the differences among these types by saying that the square, multiplied by its factor $\mathbf{a}$, is either bigger than the other two terms and is therefore equal to their sum (the last case), or is smaller than at least one of the other terms and gets added to another term to attain the largest term.

It gets easier if we look at specific examples. In the following six equations, I've conveniently picked coefficients so that $\mathbf{x} = \mathbf{5}$ is a solution and I've made repeated use of most of the coefficients:

$$3x = 15$$
$$3x^2 = 75$$
$$3x^2 = 15x$$
$$2x^2 + 15x = 125$$
$$2x^2 + 25 = 15x$$
$$4x^2 = 15x + 25$$

These equations could all be re-expressed so that they state that something equals zero, but observe that in each case we'd have to employ minus signs, jeopardizing our reluctance to use negative numbers as we worked to solve them:

$$3\mathbf{x} - 15 = 0$$
$$3\mathbf{x}^2 - 75 = 0$$
$$3\mathbf{x}^2 - 15\mathbf{x} = 0$$
$$2\mathbf{x}^2 + 15\mathbf{x} - 125 = 0$$
$$2\mathbf{x}^2 - 15\mathbf{x} + 25 = 0$$
$$4\mathbf{x}^2 - 15\mathbf{x} - 25 = 0$$

For example, a modern approach to the last of these equations $4\mathbf{x}^2 - 15\mathbf{x} - 25 = 0$ would be to factor it, giving $(4\mathbf{x} + 5)(\mathbf{x} - 5) = 0$. Clearly, when the product of two quantities is zero, one of the quantities must itself be zero. $(\mathbf{x} - 5)$ will be zero when $\mathbf{x} = 5$, but $(4\mathbf{x} + 5)$ will be zero when $\mathbf{x} = -1.25$, an unacceptable result.

Al-Khwarizmi's achievement in classifying and solving these six types of equations was a big step forward in algebra. Nitpickers might criticize the presence of one linear equation alongside five quadratics, even carping that dividing one of them, $\mathbf{ax}^2 = \mathbf{bx}$, by $\mathbf{x}$ gives the linear one, but on the whole historians of mathematics have given Al-Khwarizmi a lot more applause than raspberries. It is also apparent that al-Khwarizmi's classification set the stage for Omar Khayyam's expanded classification a few centuries later, which added the cubic equations, in all their varieties, to the simpler types of equations al-Khwarizmi had analyzed.

Like the Greeks, the Arabs enjoyed playing with numbers and discovering patterns and rules. Here is something ibn Sina found in the 10th century: for any integer n, take the first n^2 odd numbers and put them sequentially in a n-by-n square; the sum of the numbers along a diagonal of the square will be n^3, and the sum of all the numbers in the square will be n^4. Let's see what this means with actual numbers.

If the numbers 1, 3, 5, and 7 are placed in a 2-by-2 square, the diagonal sums are $1 + 7$ and $3 + 5$, which equal 2^3, and all the numbers together add up to 16, or 2^4.

If the numbers 1, 3, 5, 7, 9, 11, 13, 15, and 17 are placed in a 3-by-3 square, the diagonal sums are $1 + 9 + 17$ and $5 + 9 + 13$, which equal 3^3, and all the numbers together add up to 81, or 3^4.

If the odd numbers from 1 to 31 are placed in a 4-by-4 square, the diagonal sums are $1 + 11 + 21 + 31$ and $7 + 13 + 19 + 25$, which equal 4^3, and all the numbers together add up to 256, or 4^4.

Ibn Sina also found that the odd numbers could be placed in a triangular pattern instead of a square pattern, with an interesting result. The sum of the odd numbers in any row was the cube of the row number. $1 = 1^3 (3 + 5) = 8 = 2^3 (7 + 9 + 11) = 27 = 3^3 (13 + 15 + 17 + 19) = 64 = 4^3$.

Here is a charming, simple but subtle, result from an Arabian text: **the sum of two odd squares cannot be a square**.

Quick reflection shows us that the sum of two even squares can be a square $(6^2 + 8^2 = 10^2)$, and the sum of an odd square and an even square can be a square $(5^2 + 12^2 = 13^2)$, but why should the general formula $(a^2 + b^2 = c^2)$ fail when both a and b are odd?

Well, one useful idea to start with is that the square of an odd number is another (larger) odd number. Another useful idea is that the sum of two odd numbers is an even number.

EXERCISE

Try to prove that the square of an odd number is also odd, and that the sum of two odd numbers is an even number.

Building on these ideas, we can say that if both a and b are odd, $(a^2 + b^2)$ is even. Can this even number be a perfect square?

Even numbers that are squares are divisible by 4. (36 and 100 are examples.) We can prove this easily for all even squares:

The square root of an even square number is itself an even number, and

since even numbers are divisible by 2,

the squares of even numbers are divisible by 4.

Can the sum of two odd squares be divisible by 4? If not, that would prove the little theorem we are interested in.

An odd number has the form $(2n + 1)$, where n is an integer. When we square that, we get $(4n^2 + 4n + 1)$, which leaves a remainder of 1 when divided by 4.

A second odd number $(2m + 1)$, when squared gives $(4m^2 + 4m + 1)$. The sum of $(4n^2 + 4n + 1)$ and $(4m^2 + 4m + 1)$ would leave a remainder of 2 when divided by 4, so it's an even number that is not divisible by 4, proving the little theorem.

The Arabian mathematicians translated and studied the works of the Greek mathematicians, especially Diophantus, and they developed an equal fascination for the same sort of curious and interesting though not particularly practical relationships among numbers. The previous example of odd and even squares is a simple case. This next example, attributed to Al-Karkhi in the 11th century, is more complex. Al-Karkhi found a formula that generates solutions to the equation $x^3 + y^3 = z^2$.

This equation says simply that the sum of two cubes is a square.

You might quickly find one solution: $1^3 + 2^3 = 3^2 (x = 1, y = 2,$ and $z = 3)$.

But finding more solutions is difficult. Al-Karkhi found that if we take two rational numbers, and call them m and n, we can use them to generate appropriate x, y, and z terms.

$$x = \frac{n^2}{1 + m^3} \quad y = mx \quad z = nx$$

Observe that x is calculated based on m and n, and then y and z are calculated based on x. If $m = 2$ and $n = 3$, then $x = 1$, and we get the simple solution mentioned earlier.

PROBLEM 9.7   Solve the equation $x^3 + y^3 = z^2$ when $m = 1$ and $n = 4$.

PROBLEM 9.8   Solve the equation $x^3 + y^3 = z^2$ when $m = 3$ and $n = 4$.

PROBLEM 9.9   Solve the equation $x^3 + y^3 = z^2$ when $m = 3$ and $n = 14$.

PROBLEM 9.10   Solve the equation $x^3 + y^3 = z^2$ when $m = 5$ and $n = 42$.

EXERCISE

Prove algebraically that Al-Karkhi's method always works.

Al-Karkhi found something else charming with cubes and squares. Observe the following successive sums:

$$1+8=9$$
$$1+8+27=36$$
$$1+8+27+64=100$$
$$1+8+27+64+125=225$$
$$1+8+27+64+125+216=441$$

These sums can be restated as follows:

$$1^3 + 2^3 = (1+2)^2$$
$$1^3 + 2^3 + 3^3 = (1+2+3)^2$$
$$1^3 + 2^3 + 3^3 + 4^3 = (1+2+3+4)^2$$
$$1^3 + 2^3 + 3^3 + 4^3 + 5^3 = (1+2+3+4+5)^2$$
$$1^3 + 2^3 + 3^3 + 4^3 + 5^3 + 6^3 = (1+2+3+4+5+6)^2$$

Al-Karkhi's generalization is that: *for the first n integers, the sum of their cubes is equal to the square of their sum.*

Here is a simple Arabian extension of another Greek idea. You may recall the classification of numbers as *perfect, abundant,* or *deficient,* based on the sums of their divisors. 6 is perfect, because $1 + 2 + 3 = 6$. 18 is abundant because $1 + 2 + 3 + 6 + 9$ is more than 18. The Arabs said two numbers were *balanced* if the sums of their divisors were the same. 85 and 57 are balanced because the divisors of 85 are 1, 5, and 17, and the divisors of 57 are 1, 3, and 19, both summing to 23.

How did the Arabs find 85 and 57? They looked for even numbers that could be split into two prime numbers multiple ways. $22 = 5 + 17$. Also, $22 = 3 + 19$.

Then, $5 \times 17 = 85$ and $3 \times 19 = 57$.

So, 85 and 57 were the balanced numbers.

PROBLEM 9.11   **Find three balanced numbers whose divisors add up to 41.**

PROBLEM 9.12   **Find six balanced numbers whose divisors add up to 67.**

Finally, let's look at a practical application of geometry from around the year AD 1000. Figure 9.14 shows a trigonometric approach to determining the height of a

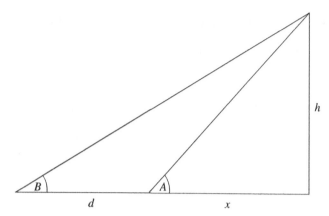

FIGURE 9.14   Computing the height of a mountain by trigonometry.

distant mountain. The height is h, and an observer on flat ground x yards away from the base of the mountain would see the very top of the mountain at an angle of elevation of $A°$. But if he didn't know the value of x (it was far away, and the terrain was difficult to traverse), he could at least say that the tangent of angle A was $\dfrac{h}{x}$. He could, however, cleverly make a second observation.

If our observer walked a measured distance further away from the mountain, d, he would then see the very top of the mountain at a new angle of elevation, $B°$. And he could say the tangent of B was $\dfrac{h}{d+x}$. Since d is known, and Arab mathematicians compiled accurate tables of sines and cosines (and since tangents are sines divided by cosines), the problem could be simplified by substitutions and solved.

Essentially, we have two formulas to express h: it is $x \cdot \tan(A)$ and $(d+x) \cdot \tan(B)$. A little algebraic manipulation leads to an expression for x: $x = \dfrac{d \cdot \tan(B)}{\tan(A) - \tan(B)}$. Since d was paced off from one observation post to another, and the tangents are numbers that can be looked up on a table, x is computable, and therefore h is also.

About 1000 years ago, a Persian scholar named al-Karaji figured out a way to solve this problem without trigonometry. He just used similar triangles and the Pythagorean theorem. In Figure 9.15, the height of the mountain, h, is one side of a right triangle with the relationship $h^2 + x^2 = y^2$, where initially h, x, and y are all unknown. A vertical line, v, is drawn at the site where angle of elevation A was measured previously, and d is the (known) distance to the observation point where angle of elevation B was measured.

To generate useful similar triangles, a line is drawn parallel to y by superimposing a copy of angle A at angle B, and another vertical line is drawn near angle B. This vertical line has two parts, with m on top of n. The small part of line d to the left of $(m+n)$ is called s, and t is the hypotenuse of the little triangle with s as its base.

There are some useful equal angles in the diagram now. The angles marked 1 are alternate interior angles between parallel line segments t and y. The three angles marked 2 are equal because two of them are vertical angles and two of them are

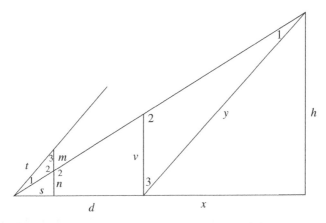

FIGURE 9.15   Computing the height of a mountain by geometry.

corresponding angles on the same side of parallel lines n and v. The angles marked 3 are equal because they occur in triangles that already have angles 1 and 2.

Now, al-Karaji saw that the ratio $\frac{s}{n}$ had to match the ratio $\frac{d}{v}$, and since s, n, and d could be measured, v could be calculated. Then, since v and y were in a triangle similar to the triangle with m and t, and since m and t could be measured, y could be calculated. Finding y was critical because $\frac{t}{s} = \frac{y}{x}$. At this point, there were two ways to confirm the value of h: the Pythagorean theorem with x, y, and h and the ratio of $\frac{m+n}{h}$.

When al-Karaji wasn't busy surveying, he found a way to construct a circle that was any desired unit fraction of a given circle, one half the size, one third, one fourth, and so on. He, like many Arabian mathematicians, dabbled in various areas of math, astronomy, geography, and medicine, as all scholarly pursuits were regarded as interconnected and insights in one intellectual field often led to understanding in others.

Suggestions for book or Internet research	
History, archeology	Abbasid Empire
	Umayyad Empire
Religion, culture	The House of Wisdom
Mathematicians	Abu Kamil
	Abu Sahl
	al-Khwarizmi
	al-Samawal
	Al-Kharki
	Omar Khayyam
Mathematical topics	Islamic mathematics
	Cubic equations
	Pentagon in a square
	Tiling patterns

ANSWERS TO PROBLEMS

9.1

The 7-11-16 triangle must be similar to the smaller triangles. $\dfrac{AX}{AC} = \dfrac{AC}{AB}$, or $\dfrac{AX}{7} = \dfrac{7}{16}$, so AX is $\dfrac{49}{16}$. $\dfrac{BY}{BC} = \dfrac{BC}{AB}$, or $\dfrac{BY}{11} = \dfrac{11}{16}$, so BY is $\dfrac{121}{16}$.

In triangle ABC, $7^2 + 11^2 = 49 + 121 = 170$.

Thabit ibn Qurra's calculation also gives 170: $(AX + BY) \cdot AB = 170$.

9.3

The formula for p, $3 \cdot 2^{n-1} - 1$, gives $3 \cdot 2^{3-1} - 1$, or $3 \cdot 2^2 - 1$, or 11. So far so good.

The formula for q, $3 \cdot 2^n - 1$, gives $3 \cdot 2^3 - 1$, or 23. Still good. As long as we generate prime numbers, the formula should work.

The formula for r, $9 \cdot 2^{2n-1} - 1$, gives $9 \cdot 2^{6-1} - 1$, or $9 \cdot 2^5 - 1$, or 287, but 287 is $7 \cdot 41$ and is therefore not a prime number.

9.5

Algebraically, we take half of the 28, 14, and square it to get 196.

We add 196 to both sides:

$$x^2 + 28x + 196 = 400$$

Now we take the square root of both sides:

$$(x + 14) = 20$$

So, $x = 6$.

Geometrically, we draw a square (x^2) in the middle, break the $28x$ into four pieces, and put them around the square, and add four 7×7 squares in the corners (to complete the square).

The sides of the big square are now $(x + 14)$ and also $\sqrt{400}$.

9.7

$$x = \frac{16}{2} = 8$$

$$y = 1 \cdot 8 = 8$$

$$z = 4 \cdot 8 = 32$$

$$8^3 + 8^3 = 32^2$$

$$512 + 512 = 1024$$

9.9

$$x = \frac{14^2}{1+3^3} = \frac{196}{28} = 7$$

$$y = 3 \cdot 7 = 21$$

$$z = 14 \cdot 7 = 98$$

$$7^3 + 21^3 = 98^2$$

$$343 + 9621 = 9604$$

9.11

$$40 = (3+37) = (11+29) = (17+23)$$

$$3 \times 37 = 111$$

$$11 \times 29 = 319$$

$$17 \times 23 = 391$$

111, 319, and 391 are the balanced numbers.
The divisors of 111 are 1, 3, and 37.
The divisors of 319 are 1, 11, and 29.
The divisors of 391 are 1, 17, and 23.

APPENDIX A: SUGGESTIONS FOR EXERCISES

ANCIENT EGYPTIAN MATHEMATICS

Page 12. This can be demonstrated with a table and an inductive proof. Start with the sums for the integers between 1 and 7.

Number	Powers of 2 to be added		
1	1		
2		2	
3	1	2	
4			4
5	1		4
6		2	4
7	1	2	4

The next whole number is 8, and then the numbers from 9 to 15 are the sums of 8 and the numbers on the previous table.

Number	Powers of 2 to be added		
8			8
9	1		8
10		2	8
11	1	2	8

(Continued)

An Introduction to the Early Development of Mathematics, First Edition. Michael K. J. Goodman.
© 2016 John Wiley & Sons, Inc. Published 2016 by John Wiley & Sons, Inc.

Number	Powers of 2 to be added			
12			4	8
13	1		4	8
14		2	4	8
15	1	2	4	8

The next whole number is 16, and then the numbers from 17 to 31 are the sums of 16 and the numbers on the previous two tables.

Number	Powers of 2 to be added				
16					16
17	1				16
18		2			16
19	1	2			16
20			4		16
21	1		4		16
22		2	4		16
23	1	2	4		16
24				8	16
25	1			8	16
26		2		8	16
27	1	2		8	16
28			4	8	16
29	1		4	8	16
30		2	4	8	16
31	1	2	4	8	16

This process can be repeated forever.

Page 12. The argument focuses on the doubling in both methods and where the odd numbers occur in the second method. Consider the problem in the text and some variations of it.

The problem 35×12 resulted in these columns:

12	35
6	70
3	140
1	280

The product of the two numbers in the first row, 12×35, is 420. The product of the two numbers in the second row, 6×70, is also 420. This is no surprise: since we doubled one factor and cut the other in half, naturally we get the same product. We've lost nothing. In the third row, 3×140 is also 420. Again, we've doubled one factor and cut the other in half, so naturally we've lost nothing. In the fourth row, something different happened. Because 3 is an odd number, we lost 140 (the number opposite 3 in the row) when we went from the third row to the fourth row, and the product of the numbers in the fourth row is 280, 140 less than 420.

The final answer requires adding the numbers opposite 1 (an odd number) on the final row, to the numbers opposite all the odd numbers above the final row (where quantities were lost when cutting in half diminished the product).

Consider the similar problem 35×13. The columns will be:

13	35
6	70
3	140
1	280

The product of the two numbers in the first row here is 455. The product of the two numbers in the second row is 420. Cutting 13 in half and throwing away the remainder had the effect of diminishing the product by the number opposite 13 on the first row, 35. You might say we split 13 into 6 and 7, and kept the 6, discarding the 7. Going from 6 to 3 lost nothing, since cutting 6 in half involved no remainder. But, as we saw before, going from 3 to 1 lost 140. The sum of the bottom number (280, opposite the odd number 1) and the numbers that were lost along the way because they were opposite odd numbers (35 opposite 13, and 140 opposite 3) is the final answer.

Consider 35×14. The columns will be as follows:

14	35
7	70
3	140
1	280

The product of the two numbers in the first row is 490. The product of the two numbers in the second row is also 490. Cutting 7 in half and throwing away the remainder diminishes the product of the next row by the number opposite 7, which is 70. And as we saw before, going from 3 to 1 lost 140. The sum of the bottom number (280, opposite the odd number 1) and the numbers that were lost along the way because they were opposite odd numbers (70 opposite 7, and 140 opposite 3) is the final answer.

Use the same technique to analyze 35×15 and 35×16. In 35×16, the product of the two numbers on every row will be 560, and the only odd number will be 1, so no halving operations will result in the loss of any part of the product.

Page 13. A multiplication can be thought of as a rectangular array. If you had 420 coins, you could arrange them in 12 rows of 35 each. If you rotated the array 90° it would appear you had 35 rows with 12 coins per row. This confirms that the order doesn't matter when we multiply two numbers.

Page 17. $\dfrac{11}{18} = \dfrac{22}{36}$

Factor pairs for 36 are as follows:

1	36
2	18
3	12
4	9
6	6

$$12+9+1=22$$

$$\frac{12}{36}+\frac{9}{36}+\frac{1}{36}=\frac{22}{36}$$

$$\frac{1}{3}+\frac{1}{4}+\frac{1}{36}=\frac{11}{18}$$

Page 17. $\dfrac{11}{18}=\dfrac{44}{72}$

Factor pairs for 72 are as follows:

1	72
2	36
3	24
4	18
6	12
8	9

$$18+12+9+4+1=44$$

$$\frac{18}{72}+\frac{12}{72}+\frac{9}{72}+\frac{4}{72}+\frac{1}{72}=\frac{44}{72}$$

$$\frac{1}{4}+\frac{1}{6}+\frac{1}{8}+\frac{1}{18}+\frac{1}{72}=\frac{11}{18}$$

Page 22. With a guess of 12, we get $3-2=1$, so the result of the guess is 1.
Then, $\dfrac{36}{1}\times12=432$.
With a guess of 120, we get $30-20=10$, so the result of the guess is 10.
Then, $\dfrac{36}{10}\times120=432$.

Page 22. The formula to find the true value of **x** requires making a fraction, the original result divided by the result from guessing, and multiplying this fraction by the guess.
The original result is **c**. The result from the guess is **z**. The guess is **y**.
$\mathbf{c}=\left(\dfrac{a}{b}\right)\mathbf{x}$, and $\mathbf{z}=\left(\dfrac{a}{b}\right)\mathbf{y}$, so the fraction reduces to $\left(\dfrac{x}{y}\right)$

Then $\left(\dfrac{x}{y}\right)\mathbf{y} = \mathbf{x}$.

Page 23. Cut each loaf into 2 halves, making 6 halves. Each man gets a half. The final half is cut into 5 equal pieces, with each piece equaling one-tenth.

Page 28. The height of the frustum is 6 and the sides of the square base at the bottom are twice as long as the sides of the base at the top. This implies that if extended upward another 6 units, the edges of the frustum would meet at the vertex of a pyramid.

Call the pyramid we add on top of the frustum **A**, and call the frustum **F**. **A + F** make a larger pyramid, **P**. The volume of the frustum is the volume of pyramid **P** minus the volume of pyramid **A**.

For **A**, volume $= \dfrac{1}{3} \cdot 2^2 \cdot 6$, or 8.

For **P**, volume $= \dfrac{1}{3} \cdot 4^2 \cdot 12$, or 64.

The volume of **F** is $64 - 8$, or 56.

Page 35. $\dfrac{8}{17} \cdot \dfrac{6}{6} = \dfrac{48}{102}$

Factor pairs for 102 are as follows:

1	102
2	56
3	34
6	17

No group of these factors adds up to 48. $34 + 17$ is too much. $34 + 6 + 3 + 2 + 1$ is too little.

$$\frac{8}{17} \cdot \frac{20}{20} = \frac{160}{340}$$

Factor pairs for 340 are as follows:

1	340
2	170
4	85
5	68
10	34
17	20

$$85 + 68 + 5 + 2 = 160$$

$$\frac{85}{340} + \frac{68}{340} + \frac{5}{340} + \frac{2}{340} = \frac{160}{340}$$

$$\frac{1}{4} + \frac{1}{5} + \frac{1}{68} + \frac{1}{170} = \frac{8}{17}$$

ANCIENT CHINESE MATHEMATICS

Page 63. The town has a diameter of 240 yards. This makes the radius of the town 120 yards. The small right triangle at the top of Figure 4.28 therefore has a short side of 120 and a hypotenuse of 255. We can calculate that the other side is 225. This triangle, 120-225-255 is the familiar 8-15-17 right triangle expanded by a factor of 15.

The large right triangle in Figure 4.28 has a short side of 208 and a side of 390. We can calculate that the hypotenuse is 442. This triangle, 208-390-442 is the familiar 8-15-17 right triangle expanded by a factor of 26.

Page 68. Initially, in the diagram (Figure 4.30), there are three unknowns: a, b, and c. But triangle a-b-c is similar to the 5-12-13 triangle.

From $\dfrac{b}{c} = \dfrac{c}{5-a}$ we can cross multiply to say $c^2 = b(5-a)$

But we also know from the diagram that $a^2 + b^2 = c^2$.
This allows us to say that $a^2 + b^2 = b(5-a)$

$$\text{or} \quad a^2 + b^2 = 5b - ab$$

This is an equation with two unknowns, a and b, but from $\dfrac{a}{b} = \dfrac{5}{12}$ we can cross multiply to say $12a = 5b$

$$\text{or} \quad a = \frac{5}{12}b$$

This allows us to substitute and create an equation with only one unknown:

$$\left(\frac{5}{12}b\right)^2 + b^2 = 5b - \left(\frac{5}{12}b\right)b$$

This is solved for b:

$$\frac{25}{144}b^2 + b^2 = 5b - \frac{5}{12}b^2$$

$$\frac{169}{144}b^2 = 5b - \frac{60}{144}b^2$$

$$\frac{229}{144}b^2 = 5b$$

$$\frac{229}{144}b = 5$$

$$b = \frac{720}{229}$$

This allows us to compute a, from the relationship $a = \dfrac{5}{12}b$, and therefore $a = \dfrac{300}{229}$

And from the Pythagorean theorem, we can compute $c = \dfrac{780}{229} = 3\dfrac{93}{229}$

Page 71. Numbers that have a remainder of 2 when divided by 10 end in 2: 12, 22, 32, 42, and so on. Numbers that have a remainder of 3 when divided by 5 end in 8 or 3: 8, 13, 18, 23, 28, and so on. No number can simultaneously end in 2 and 3 or in 2 and 8.

Algebraically, we can say a number that has a remainder of 2 when divided by 10 has the form $10k+2$ (where k is any whole number), and a number that has a remainder of 3 when divided by 5 has the form $5m+3$ (where m is any whole number). $10k+2$ cannot equal $5m+3$. Why? Because m has to be even or odd, and both options lead to a contradiction.

If m is even, then $5m$ is a multiple of 10. This makes $5m+3$ three more than a multiple of 10, while $10k+2$ is by definition two more than a multiple of 10.

If m is odd, then $5m$ is a multiple of 10, plus 5. This makes $5m+3$ eight more than a multiple of 10, while $10k+2$ is by definition two more than a multiple of 10.

Since m must be even or odd, $5m+3$ cannot be two more than a multiple of 10.

For the second impossibility, a number that has a remainder of 3 when divided by 9 has the form $9k+3$. A number that has a remainder of 2 when divided by 6 has the form $6m+2$. We are exploring whether $9k+3$ can equal $6m+2$.

Is there a solution to $9k+3 = 6m+2$ where k and m are whole numbers?

Let's subtract 2 from both sides of the equation: $9k+1 = 6m$

Since $9k = 6k+3k$, we can break $9k+1$ into two pieces:

$$6k + (3k+1) = 6m$$

Now, since $6m$ is a multiple of 6, and $6k$ must be a multiple of 6, the only way the entire left side of the equation can be a multiple of 6 is if $(3k+1)$ is also a multiple of 6. We will see that it cannot be.

If k is even, $3k+1$ is one more than a multiple of 6.

If k is odd, $3k+1$ is four more than a multiple of 6.

BABYLONIAN MATHEMATICS

Page 93. The problem would first be restated as 175 multiplied by $\dfrac{1}{25}$. If the student had not memorized the symbol for $\dfrac{1}{25}$, he could have derived it by dividing 60 parts into 25 equal shares. He'd start with [1], which he could re-express as the equivalent [0; 60]. 60 is $25+25+10$, so he would know his answer starts [0; 2, x] with x not yet determined. The x represents the "leftover" 10 sixtieths re-expressed as 3600ths and split into 25 equal shares.

10 sixtieths is (60×10) 3600ths, which is $\dfrac{600}{3600}$, which divided into 25 shares is $\dfrac{24}{3600}$ per share. Thus, the fraction $\dfrac{1}{25}$ is [0; 2, 24].

To multiply $[2,55] \times [0; 2,24]$, he could work with the 2 and the 55 separately, before adding his results. Multiplying by 55 gives a preliminary result of $[0; (55 \times 2), (55 \times 24)]$ which then becomes $[0; 110, 1320]$. 1320 is of course too large a number, but it is equal to 22 sixties, so $[0; 110, 1320]$ turns into $[0; 110+22]$ or $[0; 132]$. Now 132 is too large a number and must be converted into 2 sixties plus 12, giving $[2; 12]$.

Multiplying $[0; 2, 24]$ by 2 gives $[0; 4, 48]$, but let's not forget that in this case multiplying by 2 is really multiplying by (2×60), so the result has to be shifted one position to the left: $[0; 4, 48]$ becomes $[4; 48]$.

The two results, $[2; 12]$ and $[4; 48]$ are combined to make $[6; 60]$, which is corrected to simply $[7]$.

Page 96. On the left the bases are 17 and 13, and the height is 14. The area formula directs us to multiply the average of the bases, 15, by 14. On the right the bases are 13 and 7, and the height is 21. The area formula directs us to multiply the average of the bases, 10, by 21.

Page 101. Substitute the expression into $x^2 + Px = Q$. Squaring x gives an expression with three terms: what was under the radical, negative P times the radical, and $P^2/4$.

Multiplying P by x gives an expression with two terms, positive P times the radical, and negative $\dfrac{P^2}{2}$. These two terms cancel out nearly all the previous terms, leaving the identity $Q = Q$.

Page 105. When $p = 7$ and $q = 4$, we get the 33-56-65 right triangle.

$$7^2 - 4^2 \quad (2)(7)(4) \quad 7^2 + 4^2$$
$$33 \qquad\quad 56 \qquad\quad 65$$

When $p = 11$ and $q = 5$, we get the 96-110-146 right triangle.

$$11^2 - 5^2 \quad (2)(11)(5) \quad 11^2 + 5^2$$
$$96 \qquad\quad 110 \qquad\quad 146$$

Page 105. Squaring $(p^2 - q^2)$, and adding it to the square of $(2pq)$, gives the square of $(p^2 + q^2)$.

$$
\begin{aligned}
& p^4 - 2p^2q^2 + q^4 \\
+\ & \quad\ 4p^2q^2 \\
\hline
& p^4 + 2p^2q^2 + q^4
\end{aligned}
$$

CLASSICAL GREEK MATHEMATICS

Page 133. The combination of two ideas makes this demonstration quite straight-forward. One idea is that vertical angles are equal, and the other idea is that there are 360° around any point.

Consider the 8 vertexes of the octagon. Extend the sides of the octagon so that at each vertex there is a pair of vertical angles. The angles within the octagon sum to 1080°. The vertical angles opposite these 8 interior angles also sum to 1080°. Around each vertex there is 360°, so around all of them together there is 2880°. Between the 8 interior angles and their vertical angle equivalents, we have accounted for 2160°, leaving 720°, which must be evenly split between the 8 exterior angles and *their* vertical angle equivalents. Therefore, the sum of the exterior angles is 360°.

Page 135. The divisors of 220 are: 1, 2, 4, 5, 10, 11, 20, 22, 44, 55, and 110. The divisors of 284 are 1, 2, 4, 71, and 142.

Page 136. An odd number squared is an odd number, so the numerator of $\dfrac{\text{odd}}{\text{even}}$ can never be exactly twice the denominator. An even number squared will be a multiple of 4, so the numerator of $\dfrac{\text{even}}{\text{odd}}$ cannot be exactly twice the denominator, which will be odd.

Page 137. Break the .999999... into pieces:

$$S = \frac{9}{10} + \frac{9}{100} + \frac{9}{1000} + \frac{9}{10000} + \cdots \text{infinitely.}$$

Now, multiply by 10:

$$10S = 9 + \frac{9}{10} + \frac{9}{100} + \frac{9}{1000} + \cdots \text{infinitely.}$$

Subtract the original equation for S from this equation, leaving $9S = 9$.

Page 139. Start by considering the most familiar shape among the Platonic solids: the cube. At any corner (vertex) of the cube, 3 squares meet. We could hardly have less than 3 squares meeting, since it is impossible to enclose a volume with less than 3 squares meeting at a vertex. And we cannot have as many as 4 squares meeting, because when 4 squares meet at a point they completely surround the point, all 360° around, and make a flat surface. So, the only way to use squares to build a Platonic solid is to use exactly 3 of them, and that's what we see when we examine a cube.

Now, there are other regular polygons that could be the faces of a Platonic solid. Equilateral triangles come to mind immediately, the smallest, simplest shape. As long as we have more than 2 equilateral triangles meeting at a vertex, we can enclose a volume, and as long as we have less than 6 equilateral triangles meeting at a vertex, we can avoid making a flat surface. So, the possibilities are 3, 4, and 5 equilateral triangles, and in fact the vertexes of a tetrahedron are sites where 3 equilateral triangles meet, the vertexes of an octahedron are sites where 4 equilateral triangles meet, and the vertexes of an icosahedron are sites where 5 equilateral triangles meet.

The next shape to consider is the regular pentagon, with 5 sides. We need more than 2 pentagons (the requirement for enclosing a volume), and we cannot have more than 3 pentagons, since more than 3 of their 108° angles would overwhelm the space around a point (which is limited to 360°). In fact, the vertexes of a dodecahedron are sites where 3 regular pentagons triangles meet.

The simultaneous requirements of enclosing a volume and not making a flat surface rule out hexagons, heptagons and all regular polygons with higher numbers of sides.

Page 139. The number of faces plus the number of vertexes equals the number of edges plus two. In the square, this is 6(faces) + 8(vertexes) = 12(edges) + 2. There is some uncertainty whether the ancient Greeks knew this, but in any case European mathematicians announced it with some fanfare roughly 2000 years later.

Page 142. Triangles ACF and AEG are similar. Both have a right angle and angles of 72° and 18°. If we call side AG of the small triangle s, we can say the hypotenuse of the big triangle is $1 + 2s$. This leads to the proportion

$$\frac{1+2s}{1} = \frac{0.5}{s}$$

When this is solved, we see that $s = \dfrac{\sqrt{5}-1}{4}$ and that $1 + 2s$ (the diagonal of the pentagon) is $\dfrac{1+\sqrt{5}}{2}$.

Page 148. Triangles AXD and BDY are similar, so $\dfrac{v}{4} = \dfrac{2}{h}$. So $v = \dfrac{8}{h}$.

Applying the Pythagorean theorem to the triangles that have r as their hypotenuses, we can say $\left(1+\dfrac{8}{h}\right)^2 + 4 = r^2$ and $1 + (2+h)^2 = r^2$.

Therefore, $1 + \dfrac{64}{h^2} + \dfrac{16}{h} + 4 = 1 + 4 + h^2 + 4h$, or simply $\dfrac{64}{h^2} + \dfrac{16}{h} = h^2 + 4h$.

Multiplying by h^2 to remove denominators gives: $64 + 16h = h^4 + 4h^3$
This can be re-written as $h^4 + 4h^3 - 16h - 64 = 0$, and this can be factored

$$\left(h^3 - 16\right) \cdot \left(h + 4\right) = 0.$$

Thus, h can be $\sqrt[3]{16}$.

Page 158. $12^3 - 10^3 = 6^3 + 8^3$
This comes from the relation $\begin{array}{l} 6^3 + 8^3 + 10^3 = 12^3 \\ (216 + 512 + 1000 = 1728) \end{array}$

EARLY HINDU MATHEMATICS

Page 174. We have to be careful to distinguish between two sets of four numbers: there are the four numbers that constitute an answer to the problem, and there are the four numbers in a starting sequence that we use to find the numbers that constitute an answer.

If the smallest number of the sequence is x and the common difference is d, then the four numbers are $x, (x+d), (x+2d)$, and $(x+3d)$. The sum of these four numbers is $4x+6d$. The squares of these numbers are: x^2, $(x^2+2dx+d^2)$, $(x^2+4dx+4d^2)$, and $(x^2+6dx+9d^2)$. The sum of these squares is $(4x^2+12dx+14d^2)$.

The critical ratio is the sum of the numbers divided by the sum of the squares, or

$$\frac{4x+6d}{4x^2+12dx+14d^2}$$

We find the four numbers that constitute an answer by multiplying the four numbers in the starting sequence by this ratio.

The results of this multiplication are: $\dfrac{x\cdot(4x+6d)}{4x^2+12dx+14d^2}$, $\dfrac{(x+d)\cdot(4x+6d)}{4x^2+12dx+14d^2}$, $\dfrac{(x+2d)\cdot(4x+6d)}{4x^2+12dx+14d^2}$, and $\dfrac{(x+3d)\cdot(4x+6d)}{4x^2+12dx+14d^2}$.

The sum of these four complicated fractions turns out to be $\dfrac{(4x+6d)^2}{4x^2+12dx+14d^2}$ because each of the numerators has the term $(4x+6d)$ in it and the sum of the four leading coefficients is also $(4x+6d)$.

So, this is the sum of the four numbers that must match the sum of their squares. How are we to conveniently calculate these squares?

We can write them as follows: $\dfrac{x^2\cdot(4x+6d)^2}{(4x^2+12dx+14d^2)^2}$, $\dfrac{(x+d)^2\cdot(4x+6d)^2}{(4x^2+12dx+14d^2)^2}$, $\dfrac{(x+2d)^2\cdot(4x+6d)^2}{(4x^2+12dx+14d^2)^2}$, and $\dfrac{(x+3d)^2\cdot(4x+6d)^2}{(4x^2+12dx+14d^2)^2}$, and while this is looking really messy we actually have already seen the clever way to combine the numerators. Each has the term $(4x+6d)^2$ in it and the sum of the four leading coefficients is something we calculated before: $(4x^2+12dx+14d^2)$. Since the 4 messy fractions have the same denominator, we can express their sum as $\dfrac{(4x^2+12dx+14d^2)\cdot(4x+6d)^2}{(4x^2+12dx+14d^2)^2}$.

Happily, the term $(4x^2+12dx+14d^2)$ can be cancelled from the numerator and denominator, leaving an expression for the sum of the squares that is exactly the same expression we found earlier for the sum of the numbers.

Page 177. Start with 5 equal squares of side 1. Place four of them in a square pattern (of side 2) and chop the fifth one up in 5 bands, each $1\times\frac{1}{5}$. Place four of these bands around the 2×2 square. (This almost makes a square—there is a "missing piece" of size $\frac{1}{5}\times\frac{1}{5}$ in one corner.) Slice the fifth band into two pieces, $\frac{1}{5}\times\frac{1}{5}$, and $\frac{1}{5}\times\frac{4}{5}$. Place the $\frac{1}{5}\times\frac{1}{5}$ piece in the empty corner, completing a square (of size $\frac{11}{5}\times\frac{11}{5}$). Take the remaining $\frac{1}{5}\times\frac{4}{5}$ piece and cut it into 22 thin strips, each $\frac{1}{5}\times\frac{2}{55}$. Place these 22 thin strips around the $\frac{11}{5}\times\frac{11}{5}$ square. It will appear we now have a $\frac{123}{55}\times\frac{123}{55}$ square, but there will be one little piece missing....and that little piece will be a square of size $\dfrac{1}{55} \times \dfrac{1}{55}$. Therefore, $\dfrac{123}{55}$ is already a very good estimate for $\sqrt{5}$, and in fact $\dfrac{123}{55}$ squared is $\dfrac{15,129}{3025}$ or 5.00132231.

We could refine this estimate further by slicing razor-thin edges off the 22 thin strips and using the edges to fill in the $\dfrac{1}{55} \times \dfrac{1}{55}$ square. The razor thin edge would be $\dfrac{1}{123 \cdot 55}$, so the final formula would be $\sqrt{5} = 1 + 1 + \dfrac{1}{5} + \dfrac{2}{55} - \dfrac{1}{123 \cdot 55} = 2.23621582$ and when this is squared, we get 5.00066119.

EARLY ARABIAN MATHEMATICS

Page 189. When $n=2$, p becomes $(3 \cdot 2) - 1$, or 5. q becomes $(3 \cdot 4) - 1$, or 11. r becomes $(9 \cdot 8) - 1$, or 71. $4 \cdot 5 \cdot 11 = 220$. $4 \cdot 71 = 284$.

Page 189. When $n=4$, p becomes $(3 \cdot 8) - 1$, or 23. q becomes $(3 \cdot 16) - 1$, or 47. r becomes $(9 \cdot 128) - 1$, or 1151. $16 \cdot 23 \cdot 47 = 17,296$. $16 \cdot 1151 = 18,416$.

Page 190. The pentagon can be divided into 3 triangles by drawing diagonals from any vertex to the vertexes it isn't connected to. Each of the 3 triangles has angles that total 180°, so the sum of all the angles in all the triangles is 540°. The 5 interior angles of the regular pentagon are equal and equally share the 540°, so they are each 108°.

Page 193. CK must be $\dfrac{r}{2}$ because $ABCDE$ was constructed symmetrically inside the square, with C at the midpoint of KL. BC is 1 because it is one of the 5 equal sides of the pentagon, but it is also the hypotenuse of triangle BKC. By the Pythagorean theorem, BK is $\dfrac{\sqrt{4-r^2}}{2}$.

AJ is $\dfrac{r-1}{2}$ because JM is r and AE is 1, and JA and EM are equal. AB is of course 1. Therefore, AJB is a right triangle, and by the Pythagorean theorem JB is $\dfrac{\sqrt{3-r^2+2r}}{2}$.

Since KB and JB together make r, the expressions can be combined to say

$$\frac{\sqrt{4-r^2}}{2} + \frac{\sqrt{3-r^2+2r}}{2} = r$$

Doubling to get rid of the fractions gives

$$\sqrt{4-r^2} + \sqrt{3-r^2+2r} = 2r$$

Subtracting the first radical gives

$$\sqrt{3-r^2+2r} = 2r - \sqrt{4-r^2}$$

Squaring both sides (to start eliminating radicals) gives

$$3 - r^2 + 2r = 4r^2 - 4r\sqrt{4-r^2} + 4 - r^2$$

Subtracting 3 from both sides, and adding r^2 to both sides gives

$$2r = 4r^2 - 4r\sqrt{4-r^2} + 1$$

Adding and subtracting to isolate the remaining radical gives

$$4r\sqrt{4-r^2} = 4r^2 - 2r + 1$$

Squaring again, to eliminate the last radical gives

$$16r^2\left(4-r^2\right) = 16r^4 - 8r^3 + 4r^2 - 8r^3 + 4r^2 - 2r + 4r^2 - 2r + 1$$

When this is simplified, what's left is a fourth-degree equation:

$$32r^4 - 16r^3 - 52r^2 - 4r + 1 = 0$$

When this is solved, the value for r is approximately 1.5745

Page 200. An odd number can be considered as one more than an even number, so it can be represented as $2m+1$, where m is a whole number. $\left(2m+1\right)^2 = 4m^2 + 4m + 1$, and this must be an odd number since it is 1 more than $2 \cdot (2m^2 + 2m)$.

Two odd numbers can be represented as $(2m+1)$ and $(2k+1)$. Their sum is $2m + 2k + 2$.

Page 201. Express $x^3 + y^3 = z^2$ in terms of n and m:

$$\left(\frac{n^2}{1+m^3}\right)^3 + \left(m \cdot \left(\frac{n^2}{1+m^3}\right)\right)^3 = \left(n \cdot \left(\frac{n^2}{1+m^3}\right)\right)^2$$

Combining terms in the numerators and expressing the denominators as powers of $(1+m^3)$ gives

$$\frac{\left(n^2\right)^3}{\left(1+m^3\right)^3} + \frac{\left(mn^2\right)^3}{\left(1+m^3\right)^3} = \frac{\left(n^3\right)^2}{\left(1+m^3\right)^2}$$

To make the denominators equal on both sides, multiply the right side of the equation by $\dfrac{1+m^3}{1+m^3}$. This leads to

$$\frac{\left(n^2\right)^3}{\left(1+m^3\right)^3} + \frac{\left(mn^2\right)^3}{\left(1+m^3\right)^3} = \frac{\left(n^3\right)^2 \cdot \left(1+m^3\right)}{\left(1+m^3\right)^3}$$

Eliminating the common denominator gives

$$\left(n^2\right)^3 + \left(mn^2\right)^3 = \left(n^3\right)^2 \cdot \left(1+m^3\right)$$

Distributing the exponents, and the terms on the right gives, finally

$$n^6 + m^3 n^6 = n^6 + n^6 m^3$$

APPENDIX B: A THEORY ABOUT BASE 60 IN BABYLONIA

An enduring mystery about ancient Babylonian mathematics is why the Babylonians did their arithmetic in base 60, the so-called sexigesimal system. Using 60 as a base stands out because the majority of all cultures we know about, both past and present, used or use base 10, a choice readily explained by the fact that we have 10 fingers and conveniently use our fingers for counting. The Egyptians had successive hieroglyphic symbols for 1, 10, 100, 1000, and so on. The Chinese used special symbols to multiply their calligraphic digits by 10, 100, 1000, and so on. The Greek and Roman number systems were base 10. What made base 60 an attractive choice for the Babylonians?

Several plausible explanations have been suggested by archeologists and mathematicians, but none of them has the benefit of archeological evidence and none of them is so persuasive and compelling that experts agree that it must be the correct explanation. Figuring out why the Babylonians used base 60 therefore remains an unsolved riddle to intrigue and occupy the minds of armchair historians.

I am an armchair historian. I do not go out to Mesopotamia and dig through the remains of ancient cities, and I cannot read the cuneiform texts that have been recovered and are accessible through the generosity of museums and universities. But I enjoy pondering the questions that arise about early cultures and early mathematical thought, and this particular mystery is a gem. Many people have speculated about it.

A widespread belief is that base 60 is a compromise between two competing number systems that came into contact, thousands of years ago in the region between the Tigris and the Euphrates rivers. One system was base 5, and the other system was

An Introduction to the Early Development of Mathematics, First Edition. Michael K. J. Goodman.
© 2016 John Wiley & Sons, Inc. Published 2016 by John Wiley & Sons, Inc.

base 12, and since $5 \times 12 = 60$, users of each system could recognize and use their familiar numbers in the new hybrid system. It sounds reasonable enough, but the big flaw in my view is that we have literally thousands of those immortal indestructible clay tablets around and none of them contains mathematics done in either base 5 or base 12.

The lack of evidence is one strike. A second strike is that if indeed this compromise of different number bases occurred in Mesopotamia, because there were base 5 people and base 12 people coming into contact, isn't it reasonable to hypothesize that in similarly ancient societies similarly ancient compromises occurred? We see no evidence of base 60 (or any other compromise) in Egypt or China or, to the extent that it is understood, India. Perhaps Egypt, China, and India did not have as many competing populations as Mesopotamia did with various disparate ideas about the best way to count, but big unified empires and kingdoms were the end result of consolidation among subgroups, so there is really more likelihood that history repeated itself to some extent in these various regions.

A third strike is the notion of compromise to begin with. Groups that come into conflict often do not come as equals, and the superior or stronger culture asserts its dominance and establishes its norms, overriding the ideas and practices of the lesser society. The whole history of colonialism shows this, and if the modern Eurocentric model of cultural conquest from Columbus to the British Empire to American imperialism is too distasteful to ponder, then simply think about the example of the Roman Empire. The Romans borrowed selectively (Greek mythology and art come to mind) but basically imposed their ideas on everybody they conquered, and they pretty much conquered everybody, for several centuries.

So the compromise between base 5 and base 12 is not a great explanation for why the Babylonians wrote their numbers in base 60. Another frequently described explanation is that base 60 was such a good system for handling fractions, because 60 is divisible by so many small numbers, including 2, 3, 4, 5, and 6. Well, fractions are important, because things have to be divided, but this theory also has a few weaknesses.

First, are fractions and dividing more important than whole numbers and counting? We experience the world largely as whole numbers of things. When we look around, we see whole numbers of people, whole numbers of animals, and whole numbers of trees. Yes, we will of course need to divide things sometimes, but designing a system for the less frequent or less necessary process seems backward. Would you buy a car because of how well it handles at 120 mph (a speed you rarely if ever experience) or because of how well it handles at ordinary highway and neighborhood speeds? I'm sure ancient people were ordinarily more concerned with the whole numbers that arose from adding and multiplying than with the fractions that arose from dividing.

And if dividing was so important, why not 12, which is (like 60) divisible by 2, 3, 4, and 6? The only key missing number is 5, but the loss of that divisor is offset by the gain in simplicity. And other civilizations, the Egyptians prominently, found a completely different way to handle division and fractions, without disturbing their base 10 orientation.

Another set of suggestions is that base 60 is tied to the 60 seconds in a minute, the 60 minutes in an hour, the 30 days in a month, the 360° in a circle, or the approximately 360 days in a year. I find none of these convincing, and a few of them ludicrous. The conventions we now use—60 seconds and 60 minutes and 360°—are manmade numbers, agreed upon by people long after the Babylonians invented their base 60 arithmetic, so these conventions could hardly be the cause. The natural occurrences (30 days and 360 days) provoke more compelling arguments, but raise a few questions too. Why would a civilization so advanced that it could determine that the lunar cycle was somewhat less than 30 days choose to use a multiple of exactly 30 for its fundamental arithmetical unit? Why would a civilization so advanced that it could determine that the solar cycle was somewhat more than 360 days choose to use a divisor of exactly 360 for its fundamental arithmetical unit? It doesn't make any sense to me that they would, and in fact we can extend this negative argument by acknowledging that similar ancient societies, to some extent equally familiar with the moon and the sun, did not jump at the chance to base their numbers on a number that was a multiple of 30 and a divisor of 360.

Another odd explanation of base 60 is based on finger counting, not using the individual fingers so much, but instead the parts of the fingers. If you bend the fingers of one hand at the knuckles, you see 12 short regions, 3 per finger (not counting the thumb), and you can use a finger from your other hand to point distinctly at or to touch the 12 short regions. This would seem to be an explanation of why some people adopted base 12, not base 60, but the system could be extended, based on the positioning of the thumb or the fingers of the second hand, to signify how many other 12s are involved, in addition to the number between 1 and 12 indicated by the point or touch. It seems rather convoluted, but I see the glimmer of a useful idea in there anyway.

When I first began to seriously speculate about why the Babylonians adopted base 60, I assumed they, like other great ancient civilizations, were basically base 10 people who came to their base 10 system via the natural instinct of counting on their fingers. The big question was why they grouped six 10s together to make a 60.

The Babylonian system, after all, has a fundamental way to represent 10, 20, 30, 40, 50, and 60, but a different and more complicated way to represent 70, 80, 90, 100, 110 and all the multiples of 10 higher than 60. Why, I wondered, were six 10s so special, and I searched for a long time for something that occurred 6 or 60 times in nature or in Babylonian culture that could have been the inspiration for counting to 10 six times but no more than 6 times. There are 6 easily described directions: left, right, front, back, up, and down. There are 6 bright things that move in the night sky: Mercury, Venus, Mars, Jupiter, Saturn, and the moon, although these 6 are not equivalent in that the moon moves much more dramatically than the others, and Venus considerably outshines the other planets. Perhaps, I thought, in the pantheon of Babylonian gods there are 6 particularly important ones, or perhaps in the Babylonian territory there are 6 particularly important cities. Or perhaps, more dramatically, 60 gods or 60 cities.

Reading what I could about Babylonian culture and geography failed to uncover anything special about 6 (or 60), and a refutation of the idea that something very important occurs 6 (or 60) times in nature struck me one day and made so much sense that it seemed obvious and made me feel silly that I had not realized it before. If there were something in nature (6 directions, 6 planets, etc.,), then surely the other civilizations (Egypt and China) would have incorporated 60 into their number systems in a special way too. They didn't. So there was nothing in nature to search for, as an explanation.

But there was a natural limit to what numbers could be represented easily and unambiguously with the hands, when counting on the fingers.

For the numbers from 1 to 10, all we need to do is to raise or extend between 1 and 10 fingers.

For larger numbers, up to 59, the fingers of one hand can represent the number of tens, and the fingers of the other hand can represent the number between 1 and 9 that gets added to those tens. Since the Babylonian cuneiform wedges for 32, for example, consist of 3 horizontally oriented marks and 2 vertically oriented marks, we can reasonably infer that this representation derived from a hand gesture with 3 fingers from one hand extended horizontally and 2 fingers from the other hand extended vertically.

Just this simple protocol allows the big majority of the numbers between 1 and 59 to be shown easily. We see a modern analogy with the hand signals basketball referees use to alert game officials about which player committed a foul. Referees use fingers from two hands to indicate the jersey number of the transgressor, and there is a protocol widely followed that forbids using digits greater than 5 on the uniform numbers. A quick, unambiguous hand signal thus suffices.

The tricky numbers for signaling with the hands are the numbers that require the "vertical hand" to show more than 5 fingers: 16, 17, 18, 19, 26, 27, 28, 29, 36, 37, 38, 39, 46, 47, 48, 49, 56, 57, 58, and 59. But there are a variety of ways these numbers might have been shown, using the unique properties of the thumb.

The thumb, to take the most obvious idea, could represent 5 when extended horizontally, either across the base of the other fingers or away from the fingers, as in hitch-hiking. The 6 in 16 or 26 or 36 could be shown by an L-shape, made with the thumb and the first finger. 7 could be the extended thumb and two fingers up, a broader L-shape. But we don't know exactly what finger gestures were popular in ancient Mesopotamia, and other possibilities include pointing the fingers down instead of up, and bending the fingers to distinguish the numbers 1 through 4 from the numbers 6 through 9. The key point, which reflects the cuneiform arrangement so smoothly, is that up to five horizontal fingers could designate 10, 20, 30, 40, or 50, while some arrangement of vertical fingers designated the rest of the number.

Now the special status of the number 60 becomes clear: among people who cannot communicate effectively aloud, 60 is the first number that cannot be shown quickly and precisely by a single gesture using both hands.

People who cannot communicate aloud include people trading in a marketplace if they have no common language, and military scouts who have to convey information to their confederates without shouting, which would reveal their location. Let's say a marketplace buyer offers two coins for the merchandise the seller is holding. The seller is insulted by the low offer and raises 8 fingers, to indicate his preferred price.

The buyer puts his two coins down and raises 4 fingers, showing his willingness to pay more, but not the full 8 coins the seller is asking. The negotiations continue, wordlessly, until a compromise price is agreed to. In the military example, again it is a flash of fingers that has to convey a number. The scout has been sent to a position where he can secretly observe how large the enemy force is, and he needs to signal back to his group the number of soldiers he sees.

Of course, these simple examples do not involve writing the numbers down, and we can wonder why the Babylonians, when they got around to writing numbers, adopted the complicated base 60 system so thoroughly, with each successive "digit" representing 60 times the previous digit. I think the social status of the people who did the writing was the main explanation. The man in the street (or the merchant or the scout) generally dealt with smaller numbers (and was usually illiterate anyway), so how a number was represented was not especially important to him. It was aristocrats—government accountants, tax collectors, priests, and entrepreneurs who ran marketplaces and businesses—who needed to have a system of bookkeeping that was uniform and somewhat abstruse, intelligible to those within their fraternity but not so easy for people outside their clique to decipher.

And some things, once they get started, are almost impossible to stop. Who among us has not heard, as an explanation to our question about why a certain thing is done a certain way, that it's always been done like that. Once some Babylonian aristocrats started representing big numbers as combinations of multiples of 3600, added to multiples of 60, added to numbers smaller than 60, who could stop them? The maverick scribe who piped up and said *hey, why don't we represent big numbers as multiples of 100, added to multiples of 10, added to numbers smaller than 10?* might have been told: *shut up and do your job, and anyway what do you want me to do with this roomful of clay tablets I already made—rewrite them?*

There may come a day when a totally unexpected breakthrough in archeology reveals the actual reason the Babylonians chose base 60, but until then, I think our best guess is that a complex system of writing numbers evolved from a simple system of finger gestures and became entrenched because it served the needs of a powerful class within Babylonian society.

The nimble fingers of Babylonian scribes pressing wooden sticks into soft clay echoed the arithmetic of the common folk but had little to do with the everyday concerns of these simpler and poorer people. Records were kept for the administrators of cities and provinces, and if these people found it convenient to tally sacks of grains or herds of sheep or days elapsed by the sixties or the sixty-times-sixties, it was inevitable that one day merchant princes and actual kings would count their wealth and their subjects by the sixty-times-sixty-times-sixties and similar numbers ever higher.

FURTHER READING

AABOE, Asger, *Episodes from the Early History of Mathematics*, New York, Random House, 1964.

BECKMANN, Petr, *A History of Pi*, New York, Barnes & Noble, 1970.

BELL, Eric Temple, *Men of Mathematics*, New York, Simon & Schuster, 1937.

BERLINGHOFF, William, and GOUVEA, Fernando, *Math Through the Ages*, Washington DC, Mathematical Association of America, 2004.

BUNT, Lucas, JONES, Phillip, and BEDIENT, Jack, *The Historical Roots of Elementary Math*, New York, Dover, 1976.

BURTON, David, *A History of Mathematics*, New York, McGraw-Hill, 1991.

CAJORI, Florian, *A History of Mathematics*, New York, Chelsea, 1985.

DERBYSHIRE, John, *Unknown Quantity, a Real and Imaginary History of Algebra*, New York, Penguin, 2007.

DIAMOND, Jared, *The World Until Yesterday*, New York, Penguin, 2012.

DUNHAM, William, *Journey Through Genius, the Great Theorems of Mathematics*, New York, Penguin, 1990.

DUNN, Ross, *The Adventures of Ibn Battuta*, Berkeley, University of California Press, 1989.

EVES, Howard, *An Introduction to the History of Mathematics*, 3rd edition, New York, Holt, Rinehart and Winston, 1964.

EVES, Howard, *Great Moments in Mathematics Before 1650*, Washington, DC, Mathematical Association of America, 1983.

HEATH, Sir Thomas L., *Diophantus of Alexandria: A Study in the History of Greek Algebra*, 2nd edition, New York, Dover, 1964.

An Introduction to the Early Development of Mathematics, First Edition. Michael K. J. Goodman.
© 2016 John Wiley & Sons, Inc. Published 2016 by John Wiley & Sons, Inc.

HODGKIN, Luke, *A History of Mathematics from Mesopotamia to Modernity*, New York, Oxford University Press, 2005.

IFRAH, Georges, *The Universal History of Numbers*, Hoboken, NJ, John Wiley & Sons, Inc., 2000.

JOSEPH, George Gheverghese, *The Crest of the Peacock: Non-European Roots of Mathematics*, New York, Penguin, 1992.

KATZ, Victor, *A History of Mathematics*, New York, Harper Collins, 1993.

KATZ, Victor, editor, *The Mathematics of Egypt, Mesopotamia, China, India and Islam*, Princeton, NJ, Princeton University Press, 2007.

LEWINTER, Martin, and WIDULSKI, William, *The Saga of Mathematics: A Brief History*, Upper Saddle River, NJ, Prentice Hall, 2001.

McLEISH, John, *The Story of Numbers*, New York, Fawcett Columbine, 1992.

MOTZ, Lloyd, and WEAVER, Jefferson Hane, *The Story of Mathematics*, New York, Plenum Press, 1993.

MURDOCK, George Peter, *Our Primitive Contemporaries*, New York, Macmillan, 1934.

NEUGEBAUER, Otto, *The Exact Sciences in Antiquity*, 2nd edition, Providence Rhode Island, Brown University Press, 1957.

ROBSON, Eleanor, *Mathematics in Ancient Iraq – A Social History*, Princeton, NJ, Princeton University Press, 2008.

RUDMAN, Peter, *How Mathematics Happened – The First 50,000 Years*, Amherst, New York, Prometheus Books, 2007.

SILVERBERG, Robert, *The Man Who Found Nineveh*, New York, Holt, Rinehart and Winston, 1964.

SUZUKI, Jeff, *A History of Mathematics*, Upper Saddle River, NJ, Prentice Hall, 2002.

SUZUKI, Jeff, *Mathematics in Historical Context*, Washington DC, Mathematical Association of America, 2009.

SWETZ, Frank J., editor, *From Five Fingers to Infinity: A Journey Through the History of Mathematics*, Peru, IL, Open Court Publishing, 1994.

VAN BRUMMELEN, Glen, *The Mathematics of the Heavens and the Earth*, Princeton, NJ, Princeton University Press, 2009.

RECOMMENDED WEBSITES AND VIDEOS

For a solid, general introduction to the history of mathematics, you simply cannot do better than to start at www.storyofmathematics.com. It contains dozens of clear, informative and interesting pages that can be accessed by subject or read in chronological sequence. There is an excellent list of mathematicians and their achievements, and the "sources" tab links to additional excellent material. The most useful link is to a chronology of mathematics created by Scotland's University of St. Andrews, which itself opens up a world of articles, explanations and further links. That is, www.groups.dcs.st-and.ac.uk/-history/Chronology/full.html.

http://nrich.maths.org/1173 takes any standard fraction you input and immediately calculates an equivalent sum of Egyptian unit fractions, in several ways.

www.aina.org/books presents a long work, *Everyday Life in Babylonia and Assyria*, by H. W. F. Saggs, and it includes an excellent chapter about Babylonian scribes.

gwydir.demon.co.uk/jo/numbers/Babylon instantly converts our numbers into base 60 cuneiform wedges.

www.math.tamu.edu provides a good short overview of Babylonian mathematics.

www.math.ubc.ca explains the famous Plimpton 322 clay tablet with Pythagorean triples.

www.cut-the-knot.org/pythagoras contains dozens of different proofs of the Pythagorean theorem.

http://edhelper.com/ChineseNumbersIntro.htm provides a good introduction to Chinese calligraphic numbers.

An Introduction to the Early Development of Mathematics, First Edition. Michael K. J. Goodman.
© 2016 John Wiley & Sons, Inc. Published 2016 by John Wiley & Sons, Inc.

http://ncm.gu.se/pdf/namnaren/4548_90_1.pdf explains addition with Chinese bamboo rods.

On www.youtube.com, you can find hundreds of interesting videos about the history of math, or about history in general with some reference to how math was a part of history, including short lessons, long lectures, multipart series, and academic courses. A wonderful lightweight introductory series is John Green's **Crash Course World History**, which features The Agricultural Revolution, The Indus Valley Civilization, Mesopotamia, and Ancient Egypt in its first four installments. For speculations about numbers and counting in even earlier times, **The Dawn of Numbers** is a charming cartoon in the **MyWhyU** pre-algebra series.

David Neiman's outstanding lectures about **The Cradles of Civilization** provide a far deeper introduction (www.drdavidneiman.com). Naturally, the emphasis is history, but the wide-ranging curriculum includes all the applications of math to agriculture, commerce, astronomy, and government.

Norman J. Wildberger of Australia's University of New South Wales has a spectacular and comprehensive lecture series on the **History of Mathematics**.

There are many good short explanations of topics in Egyptian math, including **Ancient Egyptian Number System** by Tom Jaeger, **Ancient Egyptian Form of Multiplication** by Mrmaisonet, and **Egyptian Multiplication** by American Public University.

Mesopotamia: How Did Writing Begin? by Tony Sagona is a lecture that contains excellent information about cuneiform numbers. Shorter videos on this topic include: **Treasures of UCLA Library** (part 3), **Cuneiform Tablets: Ancient Writing Comes to Life**, and **Hittite Cuneiform Tablet**.

Greek mathematics is well described in a number of full-length documentary TV shows. There is a brilliant episode of the Nova series on PBS: **The Infinite Secrets of Archimedes**. There is an equally wonderful episode of the History Channel's Modern Marvels series: **Heron of Alexandria: Ancient Discoveries**. A British group called "Numberphile" has produced many good short videos, and a favorite of mine is **5 Platonic Solids**.

Ancient Chinese math is portrayed magnificently in **The Ten Classics of Ancient Chinese Mathematics**, Joseph Dauben's 93 minute lecture, presented by aaaricuny, the Asian American/Asian Research Institute of the City University of New York.

Hindu math is the focus of **The Story of Numbers (0 and 1)** narrated by Terry Jones for the me2prophet series.

Islamic Geometric Design is a fascinating online course presented by Eric Broug.

Another remarkable episode of PBS's Nova series explores the challenges faced by archeologists: **Cracking the Maya Code** follows the progress of modern scholars reading the texts and numbers of a vanished civilization.

INDEX

An Introduction to the Early Development of Mathematics, First Edition. Michael K. J. Goodman.
© 2016 John Wiley & Sons, Inc. Published 2016 by John Wiley & Sons, Inc.

Printed and bound by CPI Group (UK) Ltd, Croydon, CR0 4YY

27/10/2024

14580475-0002